Discrete Math

The Graphic Novel

Dr. Eric Gossett

Professor of Mathematics and Computer Science

Bethel University

Cover and interior illustrations by Lynn Hillesheim

Kendall Hunt
publishing company

Cover and interior images: © Eric Gossett

Terms and definitions throughout text from Discrete Mathematics with Proof, Eric Gossett. Copyright © 2009 by John Wiley & Sons, Inc. Reproduced with permission of John Wiley & Sons, Inc.

Kendall Hunt
publishing company

www.kendallhunt.com
Send all inquiries to:
4050 Westmark Drive
Dubuque, IA 52004-1840

Copyright © 2016 by Kendall Hunt Publishing Company

ISBN 978-1-5249-4856-6

Contents

Preface

$$\binom{n}{k} = \binom{n-1}{k-1} + \binom{n-1}{k}$$

This graphic novel was written to serve as a supplement to a normal discrete mathematics textbook. Besides being an enjoyable book to read, there are four ways in which I see this as a useful complement to a normal textbook.

1. The content sections of this book serve as short, easy to read overviews of a number of the topics in a typical discrete mathematics course. The graphic novel can highlight some of the major ideas without the need to provide all the details, nor does it need to cover all aspects of the topic. The student can read the appropriate section of the graphic novel, then read, in more depth, the material in the textbook.

2. The graphic novel can serve as a useful part of an exam review. Recent research[a] has shown that repeatedly rereading a text does not aid learning. The graphic novel has the advantages of being short and not identical to the main textbook. It can provide a quick reminder of the topics. The student can then look at some key sections of the main textbook to deepen that review.

3. The graphic novel format allows more opportunity to explore some common errors in student thinking. Mistakes can be presented and analyzed in more detail than is usually possible when writing a standard textbook. The length of the graphic novel prevents this from being done for every common error, but the ones that are discussed may help students identify other potential errors on their own.

4. There is some opportunity to model the kinds of questions that a successful student would ask.

Topics in discrete mathematics can be profitably arranged in many different orders. In fact, not all discrete mathematics textbooks use the same ordering. This graphic novel has been organized to make it flexible enough to work with this multitude of possible topic orders. The novel contains two kinds of units: chapters, which contain the story and should be read in numeric order, and content sections, with names that denote the content. The content sections can be read in any order. However, I would recommend reading *Sets, Set Properties, Logic*, and possibly *Elementary Number Theory* before reading *Proof*.

Discrete mathematics contains some fascinating topics. It also serves as a foundation for many other courses in mathematics and computer science. It is my hope that the reader finds this book helpful for mastering this important collection of topics.

---

[a]See the *References and Credits* section on page 231 for details.

Prologue

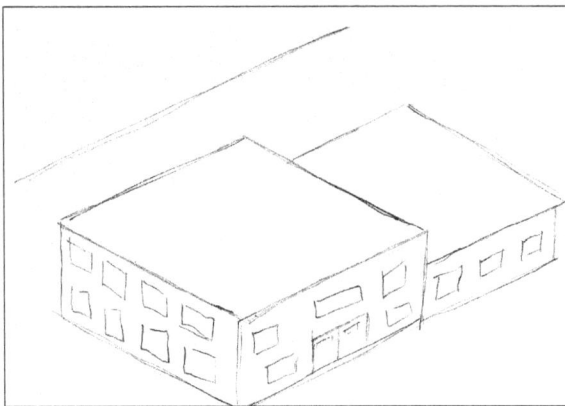

Professor Douglas, you wanted to see me. Is this a good time?

Yes, thank you for coming Isolde. Have a seat.

What was it you wanted to talk about?

I believe you are a math major who wishes to get licensed to teach. I'm seeking a junior or senior to take on an interesting teaching assignment. I immediately thought of you.

What sort of assignment is it?

I'm friends with a couple whose 14-year-old daughter is gifted in math. Her parents want to challenge her but without rushing her through advanced classes.

So they want to supplement her current classes with enriched material?

Yes. You could use material from the Discrete Math class.

Hmm...

I *am* interested. How can I contact her parents?

She is in the campus study room now if you want to meet her after the phone call.

Here is the number. The girl's name is Lily Lin: 林美恩 (lín měi ēn).

Study Room

So Lily, I will be your new tutor. My name is Isolde Gallagher.

Nice to meet you Isolde. I am looking forward to our sessions. When do we start?

We can meet here once a week. But I first need to prepare.

I can meet after school on Wednesdays. Does that work for you?

Yes. I am thinking we can study topics from Discrete Math.

Discrete math? Is that math that knows when to keep quiet?

Not exactly. You are thinking of "discreet", which means prudent. "Discrete" means "not continuous". Look at the white board.

DISCRETE
VS
DISCREET

The real numbers are continuous: if I choose any real number, r, and also pick a tiny distance, $\delta > 0$, then no matter how small I make δ, I can always find another real number that is less than distance δ from r. (In fact I can find infinitely many real numbers within δ of r.) A discrete set of numbers does not have this property: for any number n in the set, I can always choose δ small enough so that there is no other element in the set that is less than δ from n.

For example, consider the finite set $\{1, 2, 3, \pi, 4, 4.1\}$. No number in this set is within $\delta = .005$ of any other number in that set. Every finite set (of numbers or of any other kind of elements) is discrete because we can separate the elements. The set of real numbers does not have this separation property.

There are also infinite sets that are discrete. The most important examples are the natural numbers $\mathbb{N} = \{0, 1, 2, 3, \ldots\}$ and the integers $\mathbb{Z} = \{\ldots, -3, -2, -1, 0, 1, 2, 3, 4, \ldots\}$.

Any set that can be placed in one-to-one correspondence with the natural numbers is called a *countably infinite set*. (More on 1-1 correspondence and countably infinite sets in the future.)

So, discrete math is math that is primarily concerned with finite or countably infinite sets of objects.

There are lots of interesting topics we can choose from.

So where do we start?

There are several good options—you could learn how to count, or build a good foundation by starting with set theory.

Umm

…

I learned how to count in preschool.

Counting is much more involved than what you learned in preschool.

There are three topics that are foundational to pretty much all of higher mathematics: sets, logic, and the axiomatic system. From these, we can start learning how to read and construct proofs.

Logic

Sets

Constructing Proofs

Here is a list of some other topics we could explore. Look it over and email or text me with what you would like to start with first. I would suggest either set theory or counting.

This sounds fun! See you next week Isolde.

Why Should I Read A Math Book?

When I was in high school, my friends and I unconsciously held the following view about the role of math textbooks.

> The primary purpose of a math book is to provide sets of homework exercises. The student attempts to complete an exercise, then compares their answer to the one in the back of the book. If the two answers match, the student moves on to the next exercise. Otherwise, they attempt to modify their work to reproduce the correct solution. If things get desperate, they will look in the body of the chapter to find an example that looks the same, but with slightly different numbers. The student will mimic the example with the numbers from the homework exercise and hope that the correct solution will appear.
>
> At no time in this process will the student consider actually reading the body of the textbook.

This is not a helpful view of mathematics textbooks. Math books contain useful explanations about why the solution techniques actually work. By understanding the reasons for what is being done, fewer mistakes will occur. When you comprehend the material, you are able to combine ideas in new ways, perhaps solving new problems.

Many students waste time randomly making small changes in the hope that the correct solution will appear. If instead, they read and understand the textbook, they will have better insight to arrive at a solution in a systematic manner. The total time spent will often be less – with the added bonus of actually understanding the material.

A math book should be read with pencil and scratch paper at hand. When the author makes an assertion, the reader should verify the assertion or think through the reasons that justify the assertion. If that is unsuccessful, a note should be made in the book and that assertion should be addressed during office hours or in a math tutor lab. Math books should be read actively. There are few unessential sentences in a math book.

Even with classroom sessions by an accomplished teacher, the extra reinforcement achieved by a careful read of the textbook is necessary to build a solid understanding of the material. In addition, the textbook may contain material that the instructor did not have time to discuss in class. Finally, the textbook is the primary resource from which students may encounter mathematics written in a formal and detailed manner.

Lectures provide the opportunity for an interactive dialog between the student and the teacher. A textbook allows the student to consider the material at their own pace and to re-read parts they did not immediately understand. Combining the two approaches provides the student a better path to success than does either option alone.

Chapter 1

Hi Lily! Are you ready to start?

Yes, but I would like to ask a question before we do. Why did you decide to be a math major?

I want to teach math at either a high school or a middle school. In order to be a well-prepared teacher, I need to really understand mathematics well. So rather than doing an education major with a math minor, I decided to be a math major and then add the education courses that are needed for licensure.

Why do you want to be a teacher?

During my first year of high school, my math teacher was really inspiring. I found that I really liked math and wanted to be like that teacher.

$$x + y = 7$$
$$2x - y = 5$$

If you already know high school math, why do you need a bunch of university math courses?

First, to teach well, you need to know more than what you are teaching. The added depth will allow me to answer questions in a more accurate and helpful manner. Second, someday I will have a student in class who is smarter than I am. If I know more mathematics than just the course material, I will still have something to offer to that student. Also, I can say more than just "someday you will need this" – I will know why the material is important and also know where and how it will be used in the future.

That makes sense. You seem to be pretty serious about being well-prepared!

Yes, I want to be a teacher who is not just going through the motions.

Ok. I am ready to start our session now.

Great. Have a seat and let's begin.

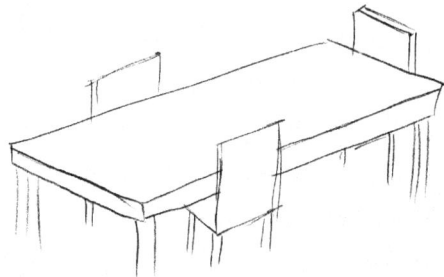

Why Are Definitions Important?

Definitions are important for several reasons. A well-chosen definition allows us to *concisely* represent a complex idea, without needing to repeat all the details every time we mention the idea. For example, it is easier to write "let p be a prime" than "let p be a positive integer that is only divisible by itself and 1". Once we know this definition, it is not difficult to unpack the details when they are needed. By memorizing the definition, we avoid being tied to a reference book or internet search engine.

Definitions also allow us to communicate more *precisely*. By having a shared, carefully crafted vocabulary, we are able to unambiguously discuss ideas. For example, the mathematical definitions for the words "positive" and "nonnegative" are not identical. Both refer to the set of integers, and both exclude negative numbers. However, "positive" excludes 0, whereas "nonnegative" includes 0. This is a subtle, but important and useful distinction. These two ideas are used often enough that they warrant formal definitions.

When you encounter a definition, you should first seek to understand the intuitive idea that motivated the definition. Once that has been accomplished, you should then memorize the definition, exactly as written. The memorization will be easier if the intuitive insight has been achieved.

The words in a definition are chosen to convey a precise meaning. Changing those words might change the meaning of the definition. Here is an example of this claim.

> **Definition** The integer a is *divisible* by the nonzero integer b if $a = bc$ for some integer c.

If the adjective "nonzero" is omitted, the modified definition would assert that zero is divisible by zero, since $0 = 0 \cdot 4$ is certainly true. But claiming that zero is a legal divisor would lead to the assertion that zero has a multiplicative inverse in the rational numbers. (This assertion follows from the general property that $a \cdot \frac{1}{a} = 1$, ensuring that $a^{-1} = \frac{1}{a}$.) But there is no rational (nor real) number c for which $0 \cdot c = 1$.

The main point of this example is that mathematical definitions tend to use very carefully chosen words. You need to pay atention to every word and phrase to fully understand the definition.

Definitions are stated (using complete sentences) by employing a common pattern: name the concept that is being defined, then list the properties it must satisfy. As part of the naming you may need to list the nature of some of the parts. Definitions are typically written in the form "name if properties", as seen in the previous example.

Sets

Today I want to introduce the initial concepts and notation used to discuss elementary set theory.

Shouldn't you say "elementary, my dear Watson"?

Actually, Sherlock Holmes never uttered that phrase. He said "elementary," and he said, "my dear Watson," but never together in the same sentence.

What we will study is called *elementary set theory* because it will be introduced informally instead of using a set of formal axioms. This approach is also called *naive set theory*.

A *set* is a collection of things or objects. The objects are called the *elements* of the set. We often use uppercase letters to name sets, and lowercase letters to name the elements in the set. So we could consider the set S with the three elements a, b, and c.

We often abbreviate this using the notation

$$S = \{a, b, c\}$$

The way that we symbolically indicate that b is an element of S is $b \in S$. We can indicate that x is *not* in S by writing $x \notin S$.

Set theory is quite versatile. We could also consider the set B consisting of the four Jane Austen books *Emma, Pride and Prejudice, Sense and Sensibility*, and *Northanger Abbey*. Although we could list the full book names in the set notation used with S, it might be more convenient to abbreviate the set as

$$B = \{e, pp, ss, na\}$$

There are four more ideas needed for our initial exploration of this topic: *union, intersection, complement*, and *subset*.

Consider the two sets $S = \{a, b, c\}$ and $T = \{a, c, d, e\}$. The *union* of S and T consists of all the elements that are elements of S or T or of both. We denote this using the symbol $\cup$ and write the result as

$$S \cup T = \{a, b, c, d, e\}$$

Notice that a and c only appear once each. Sets, by definition, do not contain duplicates. Also, sets are unordered. So $\{a, b, c\} = \{b, c, a\} = \{c, a, b\}$. Notice the $S \neq T$ since their elements are not identical.

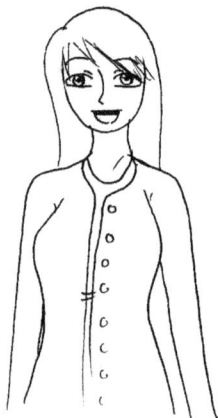

The *intersection* of two sets is a set consisting of all the elements that are in both original sets. The symbol $\cap$ is used to denote this idea. So

$$S \cap T = \{a, c\}$$

The *complement* of a set only makes sense once we define a set called the *universal set*. The universal set is the set, U, consisting of all objects potentially under consideration. So if we are discussing sets of Jane Austen novels, the universal set would consist of the four books listed previously as well as *Mansfield Park* and *Persuasion*.

The *complement* of the set R is the set, $\overline{R}$, consisting of all elements of the universal set that are *not* in R. So for the Jane Austen example

$$U = \{ss, pp, mp, e, na, p\}$$
$$B = \{e, pp, ss, na\}$$
$$\overline{B} = \{mp, p\}$$

There is a special set, called the *empty set*, denoted $\emptyset$, that does not contain any elements. Notice that $\overline{U} = \emptyset$ and $\overline{\emptyset} = U$.

Just to make sure you are understanding this Lily, please write the complements of S and T on this scratch paper. Assume $U = \{a, b, c, d, e, f\}$.

That seems pretty easy. Here is my answer.

I am curious as to why the same word is used for this idea about sets and also for praising someone (such as "I think Isolde is a great tutor").

$$\overline{S} = \{d, e, f\}$$

$$\overline{T} = \{b, f\}$$

They are not the same word. The set notion is complement. It means to complete or bring to perfection (just as Eve complemented Adam). The praise word is spelled compliment.

There is a nice way to picture the ideas we have discussed. It is called a *Venn diagram*. The diagram for the sets S and T looks like this. Notice that a and c are in both S and T, but f is in neither set.

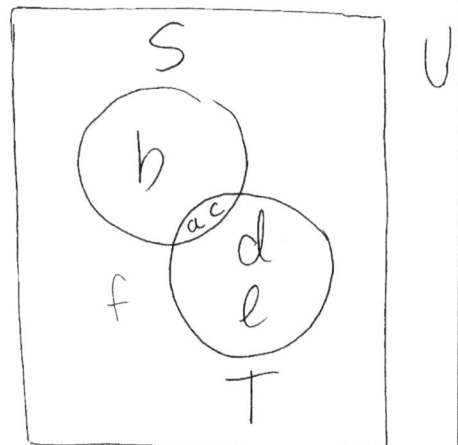

Here is a short quiz Lily: consider a set R and its complement, $\overline{R}$. What can you say about their union and their intersection?

Well ... since $\overline{R}$ contains all the elements in U that are not in R, their union must be everything. So

$$R \cup \overline{R} = U$$

If I understand the definition of $\overline{R}$, then it will have nothing in common with R, so

$$R \cap \overline{R} = \emptyset$$

The Venn diagram will look like this:

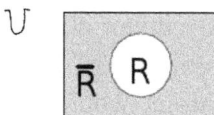

It seems you understand very well. I should mention that if $A \cap B = \emptyset$, we say that A and B are *disjoint*.

The next new idea is the notion of a subset. A *subset* A of a set B is a set whose elements are all in B. We denote this by writing $A \subseteq B$. If A contains every element of B, we may also write $A = B$ and declare the sets to be equal (identical). If we know that B contains at least one element that is not in A, we can write $A \subsetneq B$. If $A \subsetneq B$ we say that A is a *proper subset* of B. The Venn diagram for $A \subsetneq B$ looks like this:

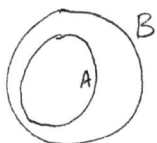

How would we write the fact that C is *not* a subset of D?

We would write $C \nsubseteq D$. I should mention that the notation $A \subset B$ is also used. Some mathematicians use $\subset$ to mean $\subsetneq$ and others use it to mean $\subseteq$. Notice that $A \cap B \subseteq A$ and $A \cap B \subseteq B$ are always true. This is true even if $A \cap B = \emptyset$ since the empty set is defined to be a subset of every set. (See exercise 9 on page 20.)

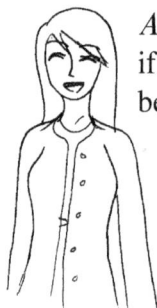

14

Here is another quiz: let $A = \{1, 3, 5, 7, 9\}$, $B = \{1, 2, 3, 4\}$, $C = \{1, 3, 4\}$, and $D = \{1, 4\}$. What are the subset relationships?

I only see $D \subsetneq C$ and $C \subsetneq B$.

You missed one: $D \subsetneq B$ is also true. We could write all three using the abbreviation $D \subsetneq C \subsetneq B$.

Do you have any questions at this point?

Nothing right now. Is there more?

There are two more ideas I want to present today. The first is a notation for the size of a set. The other is something called a power set.

If a set, S, contains n elements, we write $|S| = n$. For example, the sizes of the four sets we just examined would be $|A| = 5$, $|B| = 4$, $|C| = 3$, $|D| = 2$. Also, $|C \cup D| = 3$ and $|A \cap B| = 2$.

For a set S, its *power set*, $\mathcal{P}(S)$, is the set whose elements are all the subsets of S. Don't forget that S and $\emptyset$ are both subsets of S. If $S = \{a, b\}$ then $\mathcal{P}(S) = \{\emptyset, \{a\}, \{b\}, \{a, b\}\}$.

Lily, what is $|\emptyset|$ and what is $\mathcal{P}(\{x, y, z\})$?

I assume the empty set has size 0, since it has no elements. Is $\mathcal{P}(\{x, y, z\}) = \{\{\emptyset\}, \{x\}, \{y\}, \{z\}, \{x, y\}, \{x, z\}, \{y, z\}, \{x, y, z\}\}$?

Yes, $|\emptyset| = 0$. $\mathcal{P}(\{x, y, z\})$ is almost correct. You should have $\emptyset$ instead of $\{\emptyset\}$. They represent two different sets. The set $\emptyset$ has no elements, but the set $\{\emptyset\}$ is a set with one element, namely, the element $\emptyset$. The Venn diagrams would look like this:

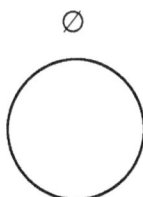

$\emptyset$

$\{\emptyset\}$

$\emptyset$

There is one final set operation I want to introduce. If A and B are two sets, then we define their *set difference*, $A - B$, to be all the elements that are in A but are not in B. For example, if $A = \{a, b, c, d\}$ and $B = \{b, c, e\}$, then $A - B = \{a, d\}$. What are the elements in $B - A$?

B contains the elements b, c, and e. But A also contains b and c, so they need to be removed. So $B - A = \{e\}$. The Venn diagrams help to see this.

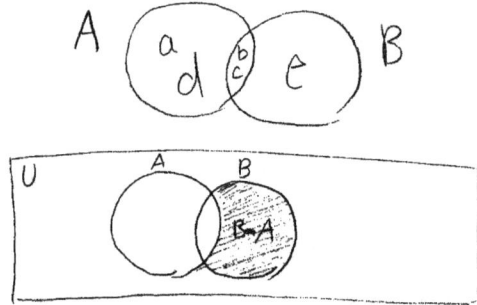

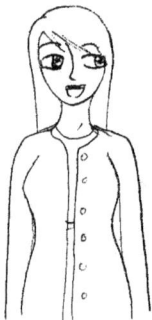

Look at the Venn diagram for $R - S$. There is a way to produce the same set using only the intersection and complement operators. Do you see how?

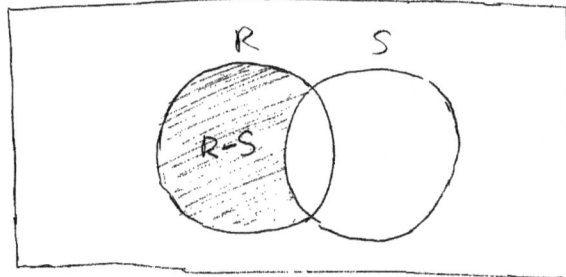

Can I get a hint?

If you take the complement of R you lose everything in R, so that is a dead end.

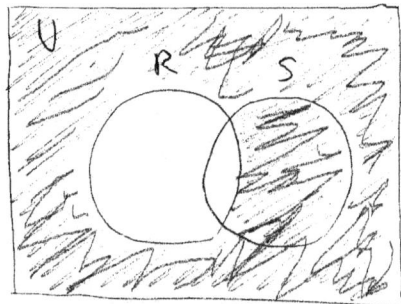

Ok. If I take the complement of S, I get this Venn diagram.

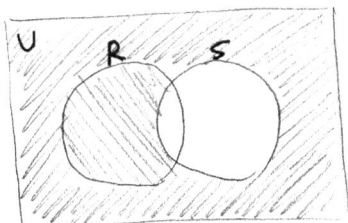

It contains what I want, but also all the elements that are in neither R nor S. If I take the intersection with R it will retain only the elements I want.

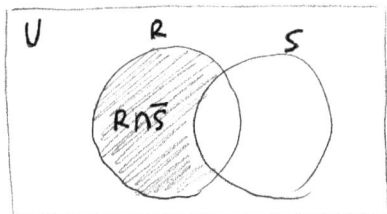

That is correct! The following theorem will be very useful when we learn about set properties.

Theorem *Set Difference*

For any two sets, A and B

$$A - B = A \cap \overline{B}$$

Which one of these is different?

I think that is enough new stuff for today. There is a lot of new notation to master, but it shows up in lots of future math classes, so it is certainly worth learning.

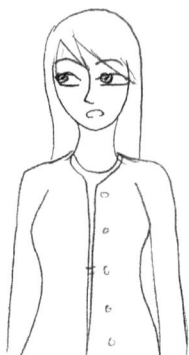

It was pretty interesting. I thought of another theorem, based on our last example: $A - B \neq B - A$, at least not always. I wonder if there are examples where they *are* equal?

Definitions

Set, Element a *set* is a collection of distinct objects, each thought of as a single entity. The set is the aggregate collection. The objects in the set are called the *elements* or *members* of the set. The potential elements are considered to be members of a set U, called the *universal set*. The set $\emptyset$, which contains no elements is called the em empty set.

The notation $x \in A$ is used to indicate that x is an element of the set A. The notation $y \notin A$ indicates that y is not an element of A.

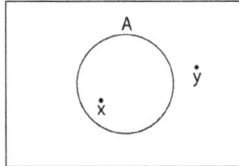

Union The *union* of the sets A and B, denoted $A \cup B$, is the set of all elements that are either in A or in B (or in both). That is, $x \in A \cup B$ if and only if either $x \in A$ or $x \in B$.

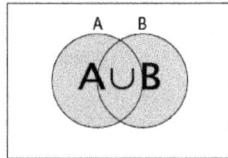

Intersection The *intersection* of the sets A and B, denoted $A \cap B$, is the set of all elements that are in both A and B. That is, $x \in A \cap B$ if and only if $x \in A$ and $x \in B$.

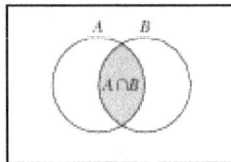

Complement The *complement*, $\overline{A}$, of the set A is the set of all elements of the universal set that are not elements of A. Thus, for every x in the universal set, $x \in \overline{A}$ if and only if $x \notin A$.

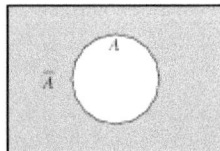

Subset The set, A, is a *subset* of a set, B, if every element of A is also an element of B. This is denoted $A \subseteq B$.

A set, A, is a *proper subset* of the set B, denoted $A \subsetneqq B$, if every element of A is an element of B, but there is at least one element of B that is *not* an element of A.

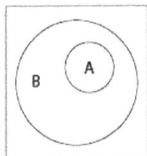

Set Difference The *set difference*, $A - B$, is the set of all elements that are in A but are not in B. That is, $x \in A - B$ if and only if $x \in A$ and $x \notin B$.

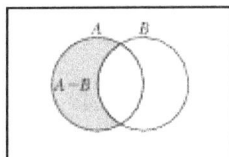

Set Difference Theorem For any two sets, A and B

$$A - B = A \cap \overline{B}$$

Disjoint The sets A and B are *disjoint* if they have no elements in common. That is, A and B are disjoint if $A \cap B = \emptyset$.

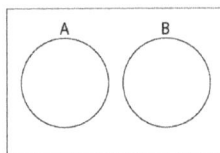

Power Set The *power set*, $\mathcal{P}(S)$, of a set S is the set whose elements are all the subsets of S. Both S and $\emptyset$ are subsets of S and are therefore elements of $\mathcal{P}(S)$.

Exercises

Solutions can be found at `http://www.mathcs.bethel.edu/~gossett/DMGN/` but solve them yourself before checking the answers.

For exercises 1 – 5, let
$$U = \{1, 2, 3, 4, 5, 6\}, \quad A = \{1, 2, 4, 5\}, \quad B = \{2, 4, 6\}, \quad C = \{1, 5, 6\}$$

1. What are $\overline{B}$, $A \cup C$, and $A \cap B$?

2. Are $(A \cup B) \cap C$ and $A \cup (B \cap C)$ the same set?

3. What are $A - B$ and $B - A$? Also, what are $|A - B|$ and $|B - A|$?

4. What is $(A \cup B) - C$? What is $\overline{(A \cap B)} - C$?

5. Is $C \subseteq A$? Is $C \subseteq (A \cup C)$? Is $(A \cap C) \subseteq C$?

6. Using (perhaps multiple times) only parentheses, union, and intersection, create an expression involving the sets A, B, and C that is distinct from $(A \cup B) \cap C$ but which does produce the same set.

7. Let $S = \{x, \{y\}\}$. What is $\mathcal{P}(S)$?

8. Let $\mathbb{N}$ represent the set $\{0, 1, 2, 3, \ldots\}$ of natural numbers. Let E represent the set of even natural numbers. One way to denote E is to write $E = \{0, 2, 4, 6, \ldots\}$.

    A better way is to use *set builder* notation and write $E = \{x \in \mathbb{N} \mid x \text{ is even}\}$, or $E = \{x \in \mathbb{N} \mid x = 2k \text{ with } k \in \mathbb{N}\}$.

    (a) Use set builder notation to write the set of natural numbers that are strictly less than 1000.

    (b) List all the elements in the set $\{x \in \mathbb{N} \mid x \text{ is a prime and } x < 30\}$. Recall that 1 is *not* a prime (neither is 0).

9. Why does it make sense to define the empty set to be a subset of every set?

Chapter 2

Good afternoon Lily. Did you have a good week?

Yes. My brother was home from the dorm last night. He is a senior majoring in computer science at this university. Do you know him? His name is Logan Lin. He said he took a course in Discrete Math a couple of years ago.

No, I haven't met him. I am a junior, so he probably took the course a year before I did.

Logan said it was a very important course for the computer science major.

```
/**
 * Set up the frame and add the tabs.
 * @param title the string to show on the frame bar
 * @param host either "localhost" or an IP of the
 * system that hosts the mysql database
 **/

public Books(String title, String host) {

    // This is a JFrame, so do all the JFrame setup

    super(title);
    addWindowListener(new WindowAdapter() {
        public void windowClosing(WindowEvent e)
        {
            mysql.closeConnection();
            System.exit(0);
        }
    });

    this.host = host;
```

DISCRETE-MATH

There were many computer science majors in the course with me. Discrete Math is also very important for math majors. It was the first course I was in that emphasized proofs as a major part of the class.

Proofs

Maybe Logan should drop by during one of our tutoring sessions. I bet you would like him.

Is she trying to play matchmaker?

Why don't we start today's session.

Set Properties

$$A \cap (B \cup \overline{C}) \cup (A \cap C) = A \cap \left((B \cup \overline{C}) \cup C\right) \quad \text{Distributivity (backwards)}$$
$$= A \cap \left(B \cup (\overline{C} \cup C)\right) \quad \text{Associativity}$$
$$= A \cap \left(B \cup (C \cup \overline{C})\right) \quad \text{Commutativity}$$
$$= A \cap (B \cup U) \quad \text{Complement}$$
$$= A \cap U \quad \text{Domination}$$
$$= A \quad \text{Identity}$$

Isolde, I heard a good math joke yesterday: Two disjoint sets dated for a few weeks, but then decided to break up because they had nothing in common.

Cute! I have not heard that one before.

Today I want to discuss some interesting properties of sets. Most of them are quite intuitive. For example, we already saw last time that if U is the universal set and $A \subseteq U$, then $A \cup \overline{A} = U$ and $A \cap \overline{A} = \emptyset$. These properties are called *complement properties*.

Two additional complement properties are $\overline{U} = \emptyset$ and $\overline{\emptyset} = U$. (Removing all elements from U leaves the empty set, and every element is an element that is not in the empty set.)

The *involution property* concerns $\overline{\overline{A}}$. Is there a simpler way to write $\overline{\overline{A}}$?

Well, ..., $\overline{A}$ is the set of elements that are not in A. The complement of that set would be the elements that are not the ones that are not in A, so I suppose that would be the elements that are back in A. That is $\overline{\overline{A}} = A$.

It is like shining a light on A. Complementing A is like flipping a switch to shine the light on $\overline{A}$. Then taking the complement again is like flipping the switch the other way, which shines the light back on A.

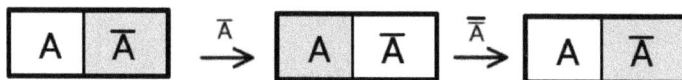

$$\boxed{A \mid \overline{A}} \xrightarrow{\overline{A}} \boxed{A \mid \overline{A}} \xrightarrow{\overline{\overline{A}}} \boxed{A \mid \overline{A}}$$

That is correct. What about the sets $A \cup A$ and $A \cap A$? (These are the *idempotence properties*. In Latin, *idem* means *same*.)

Adding elements of A to A doesn't produce anything new, since sets don't allow duplicates. So $A \cup A = A$. Also, A has every element of A in common with itself, so $A \cap A = A$. So the union and intersection of A with itself produces the same set.

A

I have nothing to add to or subtract from myself!

Here are four more very useful properties.

Domination
$A \cup U = U$ and $A \cap \emptyset = \emptyset$.

Identity
$A \cup \emptyset = A$ and $A \cap U = A$.

Commutativity
$A \cup B = B \cup A$
$A \cap B = B \cap A$.

Associativity
$(A \cup B) \cup C = A \cup (B \cup C)$
$(A \cap B) \cap C = A \cap (B \cap C)$

These all seem to be very intuitive. Are all set properties like that?

No. The next two are less intuitive. Venn diagrams are helpful to suggest their validity, but we will need more formal proofs that they are true. Here is the first:

Distributivity (∩ over ∪)
$A \cap (B \cup C) = (A \cap B) \cup (A \cap C)$
$(A \cup B) \cap C = (A \cap C) \cup (B \cap C)$

Let's think about
$A \cap (B \cup C) = (A \cap B) \cup (A \cap C)$.

The following Venn diagrams show how to derive $A \cap (B \cup C)$.

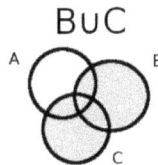

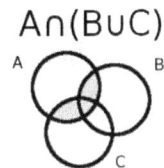

BuC

An(BuC)

The next Venn diagrams show how to derive $(A \cap B) \cup (A \cap C)$.

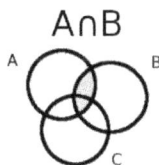

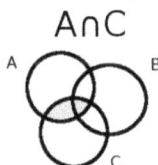

AnB

AnC

(AnB)u(AnC)

Since the diagrams for $A \cap (B \cup C)$ and $(A \cap B) \cup (A \cap C)$ are the same, we suspect that the two sets are the same.

But we don't yet know that for sure.

The Venn diagrams are pretty convincing. Why isn't that a proof?

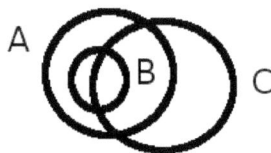

It is because we have made some assumptions about how the three sets intersect. There are other possibilities, such as these Venn diagrams. We would need to draw every possible arrangement and verify that the sets are equal in each case. And you would need to ensure that you did not miss any arrangements.

The more formal proof I will present (proofs that are even more formal can be given) uses the following strategy. To show that two sets, R and S, are equal, prove the following two assertions: (1) for every $x \in R$, show that $x \in S$ and (2) for every $y \in S$ show that $y \in R$. The first assertion shows that $R \subseteq S$ and the second shows that $S \subseteq R$. The conclusion is that $R = S$.

$$\boxed{\text{If } R \subseteq S \text{ and } S \subseteq R, \text{ then } R = S.}$$

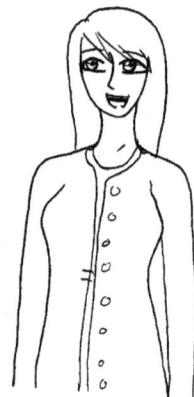

I will just prove assertion (1). Reversing the ideas will prove assertion (2).

Assume that $x \in A \cap (B \cup C)$. Focusing on the intersection, we see that $x \in A$ and $x \in (B \cup C)$. But $x \in (B \cup C)$ means that either $x \in B$ or $x \in C$ (or both). We now know that $x \in A$ and $x \in B$ or else we know that $x \in A$ and $x \in C$ (or both). Therefore $x \in (A \cap B)$ or $x \in (A \cap C)$ (or both). But that means that $x \in (A \cap B) \cup (A \cap C)$. So $A \cap (B \cup C) \subseteq (A \cap B) \cup (A \cap C)$.

Notice that this proof (actually, half a proof) does not depend on a picture for its validity. It just uses the definitions of union, intersection, and subset.

There is one last pair of set properties that we should examine. They are called the *De Morgan properties* or *De Morgan's Laws*:

$$\overline{A \cup B} = \overline{A} \cap \overline{B}$$
$$\overline{A \cap B} = \overline{A} \cup \overline{B}$$

Notice how the set operators $\cup$ and $\cap$ swap when we form the complement. A Venn diagram can suggest that these properties are valid. Instead, here is a proof of the first one.

I will again use the strategy of showing that $\overline{A \cup B} \subseteq \overline{A} \cap \overline{B}$ and $\overline{A} \cap \overline{B} \subseteq \overline{A \cup B}$ and concluding that the two sets are equal. To save effort, I will do both at once.

The acronym *iff* is an abbreviation for *if and only if*. It indicates that the step in the proof is valid in either direction.

"iff"

$$
\begin{array}{lll}
x \in \overline{A \cup B} & \text{iff} \quad x \notin A \cup B & \text{definition of } complement \\
& \text{iff} \quad (x \notin A) \text{ and } (x \notin B) & \text{definition of } union \\
& \text{iff} \quad (x \in \overline{A}) \text{ and } (x \in \overline{B}) & \text{definition of } complement \text{ (twice)} \\
& \text{iff} \quad x \in \overline{A} \cap \overline{B} & \text{definition of } intersection
\end{array}
$$

Since every element in $\overline{A \cup B}$ is also in $\overline{A} \cap \overline{B}$, we have verified that $\overline{A \cup B} \subseteq \overline{A} \cap \overline{B}$. Reading the sequence of assertions in the reverse order establishes the claim that $\overline{A} \cap \overline{B} \subseteq \overline{A \cup B}$. Consequently, $\overline{A \cup B} = \overline{A} \cap \overline{B}$.

Step 2, which appeals to the definition of union, actually uses one of the De Morgan's laws from propositional logic, which is a topic for another session.

Lily, can you make up an interesting example to illustrate De Morgan's Law for sets?

I think so. ... Let A be the the set of girls in my class with whom I am friends and let B be the set of boys in my class with whom I am friends. Then $A \cup B$ would be all the classmates who are my friends and $\overline{A \cup B}$ would be the set of classmates who are not my friends.

$\overline{A}$ would be all that are not girls who are my friends and $\overline{B}$ would be all that are not boys who are my friends. $\overline{A}$ includes some of the boys who are my friends but taking the intersection with $\overline{B}$ will eliminate them. The result is again all the students who are not my friends. (See exercise 9 on page 30.)

Now that we have a nice collection of set properties, there is another option available for proving assertions about sets.

It employs a sequence of assertions based on the properties. Here is an example, showing that $(A \cap B) \cup (A - B)$ can be simplified.

$$
\begin{aligned}
(A \cap B) \cup (A - B) &= (A \cap B) \cup (A \cap \overline{B}) && \text{Set Difference Theorem} \\
&= A \cap (B \cup \overline{B}) && \text{Distributivity (in reverse)} \\
&= A \cap U && \text{Complement} \\
&= A && \text{Identity}
\end{aligned}
$$

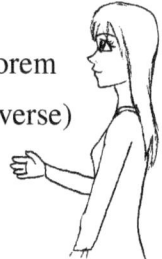

I think the hardest properties to use are the ones we tend to forget: the distributive property used from right to left and the commutative and associative properties.

Lily, see if you can use this style of proof to show that $\overline{A - B} = \overline{A} \cup B$.

Can I get a hint?

Start with one side and work your way to the other side. It is usually easier to start with the more complicated side and work towards the simpler side. Also, it is best to get rid of set difference (as I did in my example).

Some time passes.

Ok, I think I have it. It took a while to notice that the complement happens after the set difference on the left.

$$
\begin{aligned}
\overline{A - B} &= \overline{A \cap \overline{B}} && \text{Set Difference Theorem} \\
&= \overline{A} \cup \overline{\overline{B}} && \text{De Morgan} \\
&= \overline{A} \cup B && \text{Involution}
\end{aligned}
$$

That is great, Lily! I think that is enough material for today. Try some of the exercises I made up.

Set Properties

Idempotence

$A \cup A = A$

$A \cap A = A$

Domination

$A \cup U = U$

$A \cap \emptyset = \emptyset$

Associativity

$(A \cup B) \cup C = A \cup (B \cup C)$

$(A \cap B) \cap C = A \cap (B \cap C)$

Identity

$A \cup \emptyset = A$

$A \cap U = A$

Commutativity

$A \cup B = B \cup A$

$A \cap B = B \cap A$

De Morgan's Laws

$\overline{A \cup B} = \overline{A} \cap \overline{B}$

$\overline{A \cap B} = \overline{A} \cup \overline{B}$

Distributivity ($\cap$ over $\cup$)

$A \cap (B \cup C) = (A \cap B) \cup (A \cap C)$

$(A \cup B) \cap C = (A \cap C) \cup (B \cap C)$

Distributivity ($\cup$ over $\cap$)

$A \cup (B \cap C) = (A \cup B) \cap (A \cup C)$

$(A \cap B) \cup C = (A \cup C) \cap (B \cup C)$

Complement

$A \cup \overline{A} = U$

$A \cap \overline{A} = \emptyset$

Complement (continued)

$\overline{\overline{\emptyset}} = U$

$\overline{U} = \emptyset$

Involution

$\overline{\overline{A}} = A$

Set Difference

$A - B = A \cap \overline{B}$

Exercises

Solutions can be found at `http://www.mathcs.bethel.edu/~gossett/DMGN/`. Solve them yourself before checking the answers.

1. Is it always true that when $A \subseteq B$ and $B \cap C \neq \emptyset$ then $A \cap C \neq \emptyset$? If it is always true, provide a proof. If it is not always true, make up sets A, B, C that show the claim can be false.

2. Is it always true that when $A \subseteq B$ and $A \cap C \neq \emptyset$ then $B \cap C \neq \emptyset$? If it is always true, provide a proof. If it is not always true, make up sets A, B, C that show the claim can be false.

3. Prove that the subset relationship is transitive. That is, if $A \subseteq B$ and $B \subseteq C$, then $A \subseteq C$.

4. Use the set properties to provide a proof that $(A \cap C) - B = (A - B) \cap C$. Don't skip any steps.

5. Use the set properties to provide a proof that $(\overline{A} \cap \overline{B}) \cup (\overline{B} \cap \overline{C}) = (\overline{A \cap C}) - B$. Don't skip any steps.

6. Simplify the set expression $(A - \overline{A}) \cap ((A \cap B) \cup \overline{A})$. Show the set properties you use. Don't skip any steps.

7. Use the set properties to provide a proof that $(R - S) \cup (S - R) = (R \cup S) - (R \cap S)$. Don't skip any steps.

8. Is it always true that $(X - Y) - Z = X - (Y - Z)$? If it is always true, provide a proof. If it is not always true, make up sets X, Y, Z that show the claim can be false.

9. Create a set of Venn diagrams to illustrate Lilly's example of DeMorgan's Law (at the bottom of page 27).

Chapter 3

Hi Isolde! I am ready for the next lesson. And ... I brought a picture of my brother.

She is definitely trying to play matchmaker.

I am also ready. ... Your brother doesn't look familiar.

Isolde, do you ever get frustrated or discouraged when you can't solve a problem or just get stuck?

Yes, of course. However, I have recently started thinking differently about failure. I learned a new idea from one of the books assigned in an education course. – Isolde digs in her backpack and pulls out a book. – The book is named *Make it Stick: the Science of Successful Learning* by Brown, Roediger, and McDaniel. Here is a helpful passage:

... people who are helped to understand that effort and learning change the brain, and that their intellectual abilities lie to a large degree within their own control, are more likely to tackle difficult challenges and persist at them. They view failure as a sign of effort and as a turn in the road rather than as a measure of inability and the end of the road.

So instead of seeing failure as a sign that I am stupid, I see it as an opportunity to learn something new and improve.

I will think about that. Thanks for the helpful advice. Let's start now.

Logic

Today we will learn about logic—in particular, *propositional logic*. A *proposition* or a *statement* is an assertion that is either true or it is false.

For example, "Lily has a pet lizard" is a proposition.

Ok, that assertion happens to be false.

I do have a pet cat. Her name is Fluffy.

The phrase "The flowers that bloom in the spring, tra la" is *not* a proposition because it is not an assertion that is either true or false.

But "Three little maids from school are we" *would* be a proposition because the singers *are* three little maids. (I am also a fan of *The Mikado*.)

What makes propositions useful is the fact that we can glue several propositions together to make a more complex proposition. For example, the propositions "I like chocolate" and "I like cayenne peppers" can be glued together into "I like chocolate and I like cayenne peppers." We could also form "I like chocolate or I like cayenne peppers."

I could also form a new proposition by changing to "I do not like chocolate."

As you can see, the *logic connectives AND, OR, NOT* are important. We use *truth tables* to define how they work. Here is the truth table for AND. P and Q stand for any two propositions. The table defines the truth value of the compound proposition P AND Q. Each row assigns truth values to P and Q and then shows the resulting truth value of P AND Q.

P	Q	$P \wedge Q$
T	T	T
T	F	F
F	T	F
F	F	F

Do you swear to tell the truth, the whole truth, and nothing but the truth?

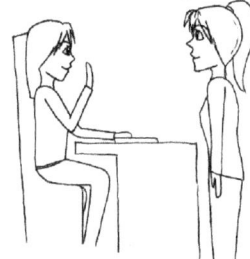

The logic connective AND is often denoted by the symbol $\wedge$.

Notice that the only way the compound proposition is true is if both component propositions are true.

The truth tables for OR and NOT are next. We use $\vee$ to denote OR and the symbol $\neg$ to denote NOT.

P	Q	$P \vee Q$
T	T	T
T	F	T
F	T	T
F	F	F

P	$\neg P$
T	F
F	T

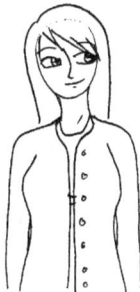

The truth table for OR may not be what you would have guessed. In casual conversation we may say something like "I will go to the store or I will stay home today." We use the "or" in that sentence as an *exclusive or*. I will do one or the other activity, but not both.

Mathematicians use an *inclusive or* for the logic connective OR. So if we say, "4 is a positive integer or 4 is an even integer," both component propositions can be true. So the default meaning of $P \vee Q$ is "P is true, or Q is true, or both are true."

"Inclusive or"

What if I want to use an exclusive or?

We denote exclusive or as XOR and use the symbol $\oplus$.

P	Q	$P \oplus Q$
T	T	F
T	F	T
F	T	T
F	F	F

I just used truth tables to define three logic operators. There is another way to use truth tables: to determine the truth values for a more complex proposition. The convention is to apply NOT before AND or OR unless parentheses are used to change the order.

P	Q	R	$\neg R$	$Q \wedge (\neg R)$	$P \vee (Q \wedge \neg R)$
T	T	T	F	F	T
T	T	F	T	T	T
T	F	T	F	F	T
T	F	F	T	F	T
F	T	T	F	F	F
F	T	F	T	T	T
F	F	T	F	F	F
F	F	F	T	F	F

Consider the compound proposition $P \vee (Q \wedge \neg R)$.

The truth table shows its value for all possible choices for the truth values of P, Q, R.

Notice how I used two intermediate columns to work towards the final column.

The next new idea is *logical equivalence*. Informally, two propositions are logically equivalent if, no matter what truth values are assigned to the logic variables, the same final truth value results. We use the symbol $\Leftrightarrow$ to indicate logical equivalence. For example,

$$\neg(P \vee Q) \Leftrightarrow \neg P \wedge \neg Q.$$

The truth table provides a proof of this claim.

P	Q	$\neg P$	$\neg Q$	$P \vee Q$	$\neg(P \vee Q)$	$\neg P \wedge \neg Q$
T	T	F	F	T	F	F
T	F	F	T	T	F	F
F	T	T	F	T	F	F
F	F	T	T	F	T	T

The last two columns have the same truth values in each position. Consequently, they are logically equivalent propositions.

Does $\neg(P \vee Q) \Leftrightarrow \neg P \wedge \neg Q$ remind you of anything?

Yes. It reminds me of De Morgan's Law for sets

$$\overline{A \cup B)} = \overline{A} \cap \overline{B}$$

That is correct! A bit later we will look at several logic properties that will be similar to the set properties. De Morgan's Law is one such property. Before that, however, there are two more logic connectives that need to be introduced.

De Morgan's Law

Sets: $\overline{A \cup B)} = \overline{A} \cap \overline{B}$

Logic: $\neg(A \vee B) \Leftrightarrow \neg A \wedge \neg B$

The new logic connectives are *implication*, denoted by →, and *biconditional*, denoted by ↔. The truth tables show how they work to glue together two component propositions into a new, compound proposition. Biconditional has the value True if and only if the two component propositions have the same truth value.

P	Q	$P \rightarrow Q$
T	T	T
T	F	F
F	T	T
F	F	T

P	Q	$P \leftrightarrow Q$
T	T	T
T	F	F
F	T	F
F	F	T

An implication, $P \rightarrow Q$, is true unless P is true and Q is false. Informally, we read this as "P implies Q," that is, if P is true, then Q must also be true. The claim would be false if P is true but Q isn't. The final two rows of the truth table are less intuitive, but they are the values that make the propositional logic definitions and properties consistent.

Ok, here is a task for you Lily. Use one or more truth tables to show that

$$\neg(P \rightarrow Q) \Leftrightarrow P \wedge \neg Q$$

This seems pretty easy. Here is my proof:

P	Q	$P \rightarrow Q$	$\neg(P \rightarrow Q)$	$\neg Q$	$P \wedge \neg Q$
T	T	T	F	F	F
T	F	F	T	T	T
F	T	T	F	F	F
F	F	T	F	T	F

Since the fourth and sixth columns are identical, the two propositions are logically equivalent.

That is correct, Lily. This provides a way to negate an implication (replacing → with ∧ and ¬).

Here is another way to look at this. If a proposition has the value True no matter how we assign truth values to its variables, we say that the proposition is a *tautology*. For example, it is easy to use a truth table to show that

$$[(P \rightarrow Q) \wedge (Q \rightarrow P)] \leftrightarrow [P \leftrightarrow Q]$$

is a tautology.

We can formally define two propositions P and Q to be logically equivalent if and only if $P \leftrightarrow Q$ is a tautology.

My head is starting to spin. Are there more logic connectives?

No, we now have all the logic connectives we need. But there are additional interesting things to do with them.

Let's think about an implication, $P \to Q$. Call it the Original implication. We can mess with negations and what is on the left and right sides of $\to$ to produce some other implications. The *derived implications* that are important are these three:

Contrapositive: $\neg Q \to \neg P$
Converse: $Q \to P$
Inverse: $\neg P \to \neg Q$

Let P represent "It is a cold day" and let Q represent "I wear a coat." Playing loose with tenses and prepositions, tell me what each of the four implications assert.

I will help you get started. The original implication can be stated as "If it is a cold day, then I wear a coat."

I think they would be something like

Contrapositive: If I don't wear a coat, then it is not a cold day.

Converse: If I wear a coat, then it is a cold day.

Inverse: If it is not a cold day, then I do not wear a coat.

Now, assume that the original assertion is actually always true for me. Which of the derived implications must also be true in all cases?

Well, if you are not wearing a coat, it can't be cold because you always wear a coat on cold days, so the contrapositive is true.
Is it possible to wear a coat if it is not a cold day? Maybe if you are totally out of shirts and need to go out to do laundry? If that is true, then the converse need not be always true. I suppose that same argument would also make the inverse false sometimes.

Correct! We can use truth tables to show that the original and the contrapositive are logically equivalent and that the converse and inverse are also logically equivalent. The original and the converse (and inverse) may or may not be logically equivalent. That needs to be checked for each original implication.

Let's finish today's session by looking at a number of very useful logical equivalences. These logic properties are very similar to the set properties, with a few more added.

Lily, can you guess what the Commutativity property would look like?

It should involve reversing the order of propositions around a logic connective. Something like $P \vee Q \Leftrightarrow Q \vee P$. I should also add $P \wedge Q \Leftrightarrow Q \wedge P$. The assertion " I will read a book or write a letter" should mean the same thing as "I will write a letter or read a book."

I am writing a letter while reading a book

You seem to understand. The two set complement properties have revised names.

Law of the Excluded Middle
$[P \vee \neg P] \Leftrightarrow T$

Law of Contradiction
$[P \wedge \neg P] \Leftrightarrow F$

Complement (Sets)
$A \cup \overline{A} = U$

$A \cap \overline{A} = \emptyset$

Here is an intuitive interpretation of the first property. The law of the excluded middle says that either P or its negation must be true—there is no third option. So either I am wearing a seat belt as I drive, or I am not. There is no partial wearing of the belt.

What about something like "Either Susie loves Joe or she doesn't." Could it be possible that she keeps bouncing back and forth with her feelings and so neither is exactly true?

She loves me, she loves me not.

In a math context, we would say that at any given moment, one of the two statements is true. She does not need to know which one.

Lily, here is a list (page 41) of some useful logic properties. We can use them to simplify propositions or to show that two propositions are logically equivalent. For example, the proposition $(P \lor Q) \land \neg(\neg P \land R)$ can be simplified.)

$$
\begin{aligned}
(P \lor Q) \land \neg(\neg P \land R) &\Leftrightarrow (P \lor Q) \land (\neg\neg P \lor \neg R) && \text{De Morgan} \\
&\Leftrightarrow (P \lor Q) \land (P \lor \neg R) && \text{Double Negation} \\
&\Leftrightarrow P \lor (Q \land \neg R) && \text{Distributivity}
\end{aligned}
$$

This is pretty much the same style in which we used the set properties. One difference is that we use logical equivalence ($\Leftrightarrow$) instead of set equality ($=$).

I think I got it.

Then you try to prove that $\neg(P \to Q) \to P$ is a tautology.

I need to show that the proposition is logically equivalent to T.

$$
\begin{aligned}
\neg(P \to Q) \to P &\Leftrightarrow (P \land \neg Q) \to P && \text{Negation of an Implication} \\
&\Leftrightarrow \neg(P \land \neg Q) \lor P && \text{Implication} \\
&\Leftrightarrow (\neg P \lor \neg\neg Q) \lor P && \text{De Morgan} \\
&\Leftrightarrow (\neg P \lor Q) \lor P && \text{Double Negation} \\
&\Leftrightarrow \neg P \lor (Q \lor P) && \text{Associativity} \\
&\Leftrightarrow \neg P \lor (P \lor Q) && \text{Commutativity} \\
&\Leftrightarrow (\neg P \lor P) \lor Q && \text{Associativity} \\
&\Leftrightarrow T \lor Q && \text{Excluded Middle} \\
&\Leftrightarrow T && \text{Domination}
\end{aligned}
$$

You did pretty well. You remembered to use Associativity. However, you forgot Commutativity in two places. The end of the proof should look like this.

$$
\begin{aligned}
&\Leftrightarrow (\neg P \lor P) \lor Q && \text{Associativity} \\
&\Leftrightarrow (P \lor \neg P) \lor Q && \text{Commutativity} \\
&\Leftrightarrow T \lor Q && \text{Excluded Middle} \\
&\Leftrightarrow Q \lor T && \text{Commutativity} \\
&\Leftrightarrow T && \text{Domination}
\end{aligned}
$$

That should be enough for today's lesson. There are some exercises for you to try. See you next week Lily.

Definitions

A *tautology* is a proposition which has the value True no matter how we assign truth values to its variables.

P	Q	$P \wedge Q$
T	T	T
T	F	F
F	T	F
F	F	F

P	Q	$P \vee Q$
T	T	T
T	F	T
F	T	T
F	F	F

P	$\neg P$
T	F
F	T

P	Q	$P \to Q$
T	T	T
T	F	F
F	T	T
F	F	T

P	Q	$P \leftrightarrow Q$
T	T	T
T	F	F
F	T	F
F	F	T

Some Useful Logical Equivalences

Idempotence
$(P \vee P) \Leftrightarrow P$
$(P \wedge P) \Leftrightarrow P$

Associativity
$[(P \vee Q) \vee R] \Leftrightarrow [P \vee (Q \vee R)]$
$[(P \wedge Q) \wedge R] \Leftrightarrow [P \wedge (Q \wedge R)]$

Commutativity
$(P \vee Q) \Leftrightarrow (Q \vee P)$
$(P \wedge Q) \Leftrightarrow (Q \wedge P)$

Distributivity ($\wedge$ over $\vee$)
$[P \wedge (Q \vee R)] \Leftrightarrow [(P \wedge Q) \vee (P \wedge R)]$
$[(P \vee Q) \wedge R] \Leftrightarrow [(P \wedge R) \vee (Q \wedge R)]$

Law of the Excluded Middle
$[P \vee \neg P] \Leftrightarrow \mathbf{T}$

Law of Double Negation (Involution)
$\neg(\neg P) \Leftrightarrow P$

Implication
$(P \to Q) \Leftrightarrow [\neg(P \wedge \neg Q)]$
$\qquad\quad \Leftrightarrow [\neg P \vee Q]$

Domination
$(P \vee T) \Leftrightarrow T$
$(P \wedge F) \Leftrightarrow F$

Identity
$(P \vee F) \Leftrightarrow P$
$(P \wedge T) \Leftrightarrow P$

De Morgan's Laws
$[\neg(P \vee Q)] \Leftrightarrow [\neg P \wedge \neg Q]$
$[\neg(P \wedge Q)] \Leftrightarrow [\neg P \vee \neg Q]$

Distributivity ($\vee$ over $\wedge$)
$[P \vee (Q \wedge R)] \Leftrightarrow [(P \vee Q) \wedge (P \vee R)]$
$[(P \wedge Q) \vee R] \Leftrightarrow [(P \vee R) \wedge (Q \vee R)]$

Law of Contradiction
$[P \wedge \neg P] \Leftrightarrow F$

Law of Addition
$[P \to (P \vee Q)] \Leftrightarrow T$

Negation Of An Implication
$[\neg(P \to Q)] \Leftrightarrow [P \wedge \neg Q]$

The Biconditional
$(P \leftrightarrow Q) \Leftrightarrow [(P \to Q) \wedge (Q \to P)]$

Exercises

Solutions can be found at `http://www.mathcs.bethel.edu/~gossett/DMGN/`.

1. Use truth tables to determine whether the following propositions are logically equivalent.

   (a) $A \wedge (B \vee \neg C)$ $A \vee (\neg B \wedge C)$

   (b) $(P \leftrightarrow Q) \vee \neg Q$ $P \vee \neg Q$

   (c) $X \rightarrow Y$ $\neg Y \rightarrow \neg X$

2. Consider the implication: "If Johnny is hungry and his mom is not home, then he eats cookies."

   (a) Create variables for the three constituent propositions.

   (b) Write the implication in symbolic form (using the variables).

   (c) Write the three derived implications, both in symbolic form, and in English.

   (d) For this example, is the converse likely to be true in all cases?

   (e) Use the useful logical equivalences on the previous page to write the negation of the original implication, in English.

3. The implication $[P \wedge (P \rightarrow Q)] \rightarrow Q$ is related to a rule for reasoning named *modus ponens*. It essentially says that if we know that P is true and we also know that the implication $P \rightarrow Q$ is true (if P is true then so is Q), then we know that Q is true. Prove that this is a tautology.

   (a) Using truth tables.

   (b) Using the table of useful logical equivalences on the previous page. Pay attention to associativity. Add parentheses where necessary. Don't forget the commutativity property.

4. Use the table of useful logical equivalences on the previous page to prove these logical equivalences.

   (a) $[P \rightarrow Q] \Leftrightarrow [(P \wedge \neg Q) \rightarrow Q]$ (proof by contradiction)

   (b) $[(P \wedge Q) \rightarrow P] \Leftrightarrow T$ (the law of simplification)

   (c) $[(A \vee B) \wedge (A \vee \neg B)] \rightarrow A \Leftrightarrow T$

Chapter 4

Isolde! I brought Logan to meet you.

Hi Isolde. I'm Lily's brother Logan. She talks about you a lot and insisted that I come meet you.

He is cuter than I guessed from his picture.

It is nice to meet you. Lily has mentioned you several times.

There is quite an age difference between the two of you. Do you have any other siblings?

No, just the two of us. Our father started grad school after I was born and our parents decided that they could only handle one child during that time. Lily was kind of like a victory crown after he finished his Ph.D.

Our mom said Logan was the practice model and I was the final product. (^_^)

Do you have any brothers or sisters Isolde?

No. I am the only child in my family. It was a bit boring on long family trips in the car.

I have a puzzle for the two of you to solve. It was publicized in Germany by Georg Christoph Lichtenberg in the 18th century.

You are in a cemetery and see a fenced-in family plot. There is a sign, with six graves in the plot. How can the sign be true?

Isolde and Logan work a while, then

This is a nasty puzzle, Lily. We can solve it if there are 8 corpses. We can indicate generations by the initial letter of the first name: A is the first generation, B is second, C is third. The people are Abigail Smith, Annie Jones, Bob and Bonnie Smith, Bill and Betty Jones, and Cathy Smith and Carrie Jones.

Abigail is the widowed mother of Bob and Bonnie, Annie is the widowed mother of Bill and Betty. Bob marries Betty and Cathy is their daughter. Bill marries Bonnie and their daughter is Carrie. Cathy and Carrie never marry. These 8 people fit the claims on the sign.

Solid single arrow means parent - child.
Dashed double arrow means spouse.

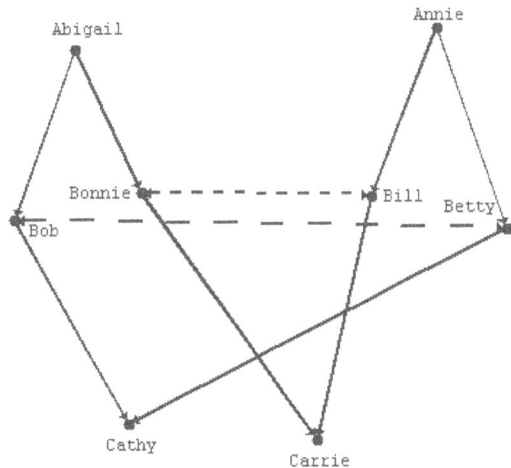

Abigail

Annie

Bonnie

Bill

Betty

Bob

Cathy

Carrie

That is good, but the puzzle can be solved with only 6 corpses. However, you will need to be a bit flexible on interpreting the final requirement. it should really read:
2 Sisters with their 2 Half-Brothers.

It is time to start the tutoring session. We will think more about this over the next week.

45

Why Should I Memorize?

You may have wondered why your professors insist that you memorize definitions, theorems, logarithm properties, values of trigonometric functions at standard angles, and other mathematical ideas. Why bother when you could find those items in a book or on the internet whenever you need them?

Here are a number of reasons why memorizing can be to your advantage.

- Memorizing basic definitions, theorems, properties, and other facts allows you to be more efficient both when reading new material and when doing assignments. If you haven't memorized ideas that are mentioned in the reading or exercise, you will need to (continually) stop and go look up the relevant idea. This wastes time and also distracts you from the main flow of what you should really be concentrating on.

- In addition to gaining efficiency, the memorized ideas aid in lowering the cognitive load when working with new material. If you are trying to master indirect proof (page 68), you will need to think much harder than necessary if you don't understand the notion of the contrapositive of an implication (page 38). That understanding will inevitably involve having memorized what the contrapositive means and what it looks like.

- Deep learning requires ideas to be attached to other ideas, creating a rich network of relationships among ideas. If you have not memorized many ideas, new ideas have nowhere to attach. If you don't have those memorized ideas to build onto, you will end with a shallow collection of unrelated ideas, each seeming to be as important as the others. Memorization enables a complex and useful network of ideas.

- If you have not memorized certain ideas, you may encounter them while reading or trying to solve an exercise, but never realize that the idea is present. Thus, instead of pausing to look up what you are missing, you may never notice that there is something you need to find. Consequently, the unmemorized information is completely unavailable for you to use, leaving you stuck and frustrated.

Memorize new ideas as early as possible: the night before an exam is the wrong time (you will mix bits and pieces of distinct ideas together into mush). Instead, start memorizing new ideas as soon as they are encountered. Keep reviewing them all through the course. The night before an exam is the time to do one last check that you have done the memorization properly and completely.

Elementary Number Theory

Theorem *The Quotient–Remainder Theorem*

Let a and b be integers with $b \neq 0$. Then there exist unique integers q and r such that $a = bq + r$ and $0 \leq r < |b|$.

Theorem *The Fundamental Theorem of Arithmetic*

Every integer n, with $n \geq 2$, can be uniquely written as a product of primes in ascending order.

Today's topic will be pretty easy—and perhaps mostly a review of things you already know. The topic is *elementary number theory*.

You mean stuff I learned in elementary school?

You *did* learn some of this in elementary school. However, in this context, the word *elementary* just means *beginning* or *rudimentary*.

Elementary number theory takes an informal, rather than a formal, axiomatic approach to the topic (see page 64 for more on the axiomatic approach).

I will begin by establishing several collections of numbers. All of the numbers we will need are part of the system of *real numbers*. (The real numbers are themselves contained inside a larger system named the *complex numbers*, but we will not need that system.)

The Real Numbers $\mathbb{R}$

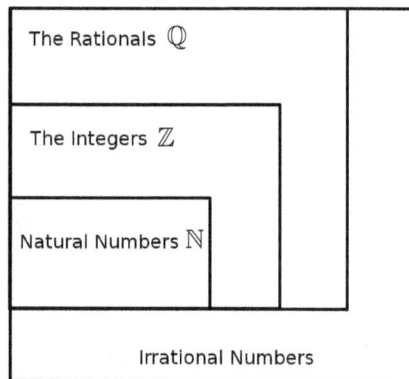

The Rationals $\mathbb{Q}$

The Integers $\mathbb{Z}$

Natural Numbers $\mathbb{N}$

Irrational Numbers

We denote the real numbers using the symbol $\mathbb{R}$. $\mathbb{R}$ is a disjoint union of the system of *rational numbers*, denoted $\mathbb{Q}$, and the *irrationals*. The rationals are the familiar set of fractions. The irrationals are all the real numbers that are *not* rational. Inside the set of rationals is the set of *integers*, denoted $\mathbb{Z}$. The integers consists of the numbers $\ldots, -4, -3, -2, -1, 0, 1, 2, 3, 4, \ldots$ The set of nonnegative integers (0, 1, 2, 3, ...) is named the *natural numbers* and denoted $\mathbb{N}$.

The reals, the rationals, and the integers are all formal systems that are defined by sets of axioms. The irrationals and the natural numbers are not systems defined by axioms.

Elementary number theory is a topic about the integers, but it is important to know the formal definition for the set of rational numbers.

Definition *The Rational Numbers*

The set of *rational numbers* is denoted by $\mathbb{Q}$, and is defined by

$$\mathbb{Q} = \left\{ \frac{p}{q} \;\middle|\; p \in \mathbb{Z}, q \in \mathbb{Z}, \text{ and } q \neq 0 \right\}$$

Read this as: "$\mathbb{Q}$ is the set of numbers that can be expressed in the form $\frac{p}{q}$, where p and q are integers and $q \neq 0$."

We can think of the integers as a subset of the rationals. For example, 4 can be written $\frac{4}{1}$.

The word *rational* emphasizes that these numbers can be thought to express a *ratio* of two integers.

The next definition will find its way into several other definitions and theorems.

Definition *Divisible*

The integer a is *divisible* by the nonzero integer b if $a = bc$ for some integer c. We denote this by $b \mid a$ and also say that b *divides* a. We call b and c *divisors* of a.

We do not discuss divisors of real numbers that are not integers because for *any* real numbers, z and $x \neq 0$, we can find a real number y such that $z = xy$. However, the integer 6 is *not* divisible by 5, so the concept of divisibility has greater significance in the set of integers.

The following definitions are all about the integers. When you encounter a term, such as *even* or *prime*, there is always an implicit assumption that we are talking about integers. There is no such thing as a prime rational number that is not an integer. There are no even irrationals.

Notice that if the integer n is composite, then we can find integers $a > 1$ and $b > 1$ such that $n = ab$.

Definition *Prime, Composite*

An integer, $p > 1$, is *prime* if its only divisors are 1 and p. An integer, $n > 1$, is *composite* if it is not prime. The number 1 is neither prime nor composite.

Definition *Even, Odd*

An integer, n, is *even* if there is an integer k such that $n = 2k$. An integer is *odd* if it is not even.

The number 0 is even since $0 = 2 \cdot 0$. If n is odd, then we can always find an integer k such that $n = 2k+1$.

Notice that the sets of even and odd numbers partition the integers into two disjoint sets.

Hint: When discussing two possibly distinct even or odd numbers, you need to recruit two letters. For example, if you want to discuss an even number a and an odd number b, write $a = 2k$ and $b = 2j + 1$.

Writing $a = 2k$ and $b = 2k+1$ implies that $b = a + 1$, which is usually not what you intend.

Lily, can you use these two definitions to prove that the sum of an even and an odd integer is odd?

I think I can. If I use your hint and assume the numbers are a and b, then

$$a + b = 2k + (2j + 1) = 2(k + j) + 1$$

But $k + j$ is just some other integer, say $k + j = m$. Since the sum can be written in the form $2m + 1$, I know it is not divisible by 2, so it must be odd.

I suppose this shows that it is helpful to remember how to look for common factors.

I am ready to introduce a very important theorem. It is one you know, but perhaps not in the form I will state it.

Theorem *The Quotient–Remainder Theorem*

Let a and b be integers with $b \neq 0$. Then there exist unique integers q and r such that $a = bq + r$ and $0 \leq r < |b|$.

The theorem states that there will always be a quotient and a remainder. The remainder is always smaller (in absolute value) than the divisor. Also, the quotient and the remainder are unique: there really is only one correct answer.

For example, $33 = 9 \cdot 3 + 6$. In the example, $a = 33$, $b = 9, q = 3$, and $r = 6$.

We can even have $b > a$. For example, to divide 12 by 45: $12 = 45 \cdot 0 + 12$.

This reminds me of third grade! My teacher called it *long division*. Your first example would be written:

$$\begin{array}{r} 3 \;\; r6 \\ 9\overline{\smash{\big)}\,33} \\ \underline{27} \\ 6 \end{array}$$

Since I learned it long ago, why is this important?

There are lots of applications. Here is an easy one: an integer b is divisible by the integer a if and only if the reminder is 0. That is, $b = aq + 0$.

Later in this session I will define the *mod* operator. Finding remainders will be an essential tool in that context.

The next ideas involve common divisors and common multiples.

Definition *Greatest Common Divisor*

Let a and b be integers that are not both 0. The *greatest common divisor* (gcd) of a and b is a positive integer d such that

- $d \mid a$ and $d \mid b$.

- If c divides both a and b, then $c \mid d$.

The gcd of a and b is denoted by $\gcd(a, b)$.

The first of the two conditions on d ensures that d is a *common divisor*. The second condition ensures that it is the largest of all common divisors. Notice that the definition of gcd is symmetric in a and b: $\gcd(a, b) = \gcd(b, a)$.

Definition *Least Common Multiple*

The *least common multiple* (lcm) of two integers a and b is a nonnegative integer, m, such that

- $a \mid m$ and $b \mid m$.

- If both a and b divide c, then $m \mid c$.

The lcm of a and b is denoted by $\text{lcm}(a, b)$.

One way to find the gcd and lcm of two integers is to find the prime factorization of both integers. The gcd contains only the common factors, the lcm contains all the factors that are not duplicated by the integers (sort of like an intersection and a union). For example

$a = 44 = 2^2 \cdot 11$
$b = 66 = 2 \cdot 3 \cdot 11$
$\gcd(44, 66) = 2 \cdot 11 = 22$
$\text{lcm}(44, 66) = 2^2 \cdot 3 \cdot 11 = 132$

Here is the key idea. Suppose a has the prime factorization $a = p^i q^j r^k$ and b has the prime factorization $b = p^u q^v r^w$, where i, j, k, u, v, w are all greater than or equal to 0. Then

$$\gcd(a, b) = p^{\min\{i,u\}} q^{\min\{j,v\}} r^{\min\{k,w\}}$$
$$\operatorname{lcm}(a, b) = p^{\max\{i,u\}} q^{\max\{j,v\}} r^{\max\{k,w\}}$$

For example, let $a = 19278675 = 3^3 \cdot 5^2 \cdot 11^0 \cdot 13^4$ and $b = 226875 = 3^1 \cdot 5^4 \cdot 11^2 \cdot 13^0$. Then

$$\gcd(a, b) = 3^1 \cdot 5^2 \cdot 11^0 \cdot 13^0 = 75$$
$$\operatorname{lcm}(a, b) = 3^3 \cdot 5^4 \cdot 11^2 \cdot 13^4 = 58317991875$$

If you use your calculator, you can verify that $75 \mid a$ and $75 \mid b$. Also, $a \mid 58317991875$ and $b \mid 58317991875$.

The gcd can be used to determine whether two integers have *any* common factors.

Definition *Relatively Prime*

Positive integers a and b are *relatively prime* if $\gcd(a, b) = 1$.

So 5 and 6 are relatively prime, but 4 and 6 are not (since $\gcd(4, 6) = 2$).

It all seems pretty familiar, except I haven't heard the phrase "relatively prime" before.

There is one more concept for today that probably will also be new for you.

Definition *$a \bmod m$*

Let m be a positive integer. Then $a \bmod m$ is the remainder when a is divided by m. The integer m is called the *modulus*.

For example, $12 \bmod 5 = 2$ and $12 \bmod 4 = 0$.

A related notion is that of congruence.

Definition *$a \equiv b \pmod{m}$*

We say that a is *congruent to* b, mod m, if m divides $a - b$. This is often expressed as: $a \equiv b \pmod{m}$ if and only if there is an integer k for which $a - b = km$.

So a and b are congruent, mod m if they have the same remainder on division by m. For example, $25 \equiv 36 \pmod{11}$.

That is a lot of definitions! I assume there is a reason why they are important?

Those ideas have been around for a long time. Even though they are mostly pretty simple, there are quite a few applications in other areas of mathematics.

One of those applications uses congruence. The set $\mathbb{Z}_n$ of integers mod n consists of all the possible remainders when dividing by the positive integer n. That is, $\mathbb{Z}_n = \{0, 1, 2, 3, \ldots, n-2, n-1\}$. We can define addition and multiplication for elements in this set by adding or multiplying in $\mathbb{Z}$ and then taking the remainder after dividing by n.

Consider, $\mathbb{Z}_5 = \{0, 1, 2, 3, 4\}$. Let $+_{\mathbb{Z}_5}$ be the addition in $\mathbb{Z}_5$ and $+_{\mathbb{Z}}$ be the familiar addition in $\mathbb{Z}$. Then for $a, b \in \mathbb{Z}_5$, $a +_{\mathbb{Z}_5} b = a +_{\mathbb{Z}} b$ mod 5. The subscripts on the $+$ signs are awkward, so we tend to omit them and just remember that we are using two different addition operators.

Using this definition in $\mathbb{Z}_5$, $2+4 = 6$ mod $5 = 1$ and $1 + 4 = 5$ mod $5 = 0$. In a similar manner, $2 \cdot 4 = 8$ mod $5 = 3$. In this way, we can create addition and multiplication tables for $\mathbb{Z}_5$. This strange number system has many of the same properties as the integers, but consists of only a finite number of elements.

+	0	1	2	3	4
0	0	1	2	3	4
1	1	2	3	4	0
2	2	3	4	0	1
3	3	4	0	1	2
4	4	0	1	2	3

$\cdot$	0	1	2	3	4
0	0	0	0	0	0
1	0	1	2	3	4
2	0	2	4	1	3
3	0	3	1	4	2
4	0	4	3	2	1

There is a theorem about the gcd that is very useful in more advanced math classes.

Theorem gcd$(a, b) = as + bt$

Let a and b be integers such that at least one is not 0. Then there are integers, s and t, such that gcd$(a, b) = as + bt$.

For example,
gcd$(242, 88) = 22 = 242 \cdot (-1) + 88 \cdot (3)$.

How do you calculate s and t?

53

The most common way is to use the extended Euclidean Algorithm, which is based on the Quotient–Remainder Theorem. It is a two-phase process that takes some time, so I think we will skip the details. You can look up the algorithm on Wikipedia if you are curious.

I will wait until I am older to learn about the extended Euclidean Algorithm. (^_^)

I should mention that the integers s and t are not unique. Notice that

$$\gcd(6, 4) = 2 = 6 \cdot 1 + 4 \cdot (-1)$$
$$= 6 \cdot 3 + 4 \cdot (-4)$$
$$= 6 \cdot 7 + 4 \cdot (-10)$$

You can learn about the extended Euclidean Algorithm when you get to be my age.

I should briefly mention the floor and ceiling functions. They turn real numbers into integers, but not by rounding.

Definition *Floor Function; Ceiling Function*

The *floor* and *ceiling* functions are defined for all real numbers x:

$$\text{floor}(x) = \lfloor x \rfloor = \text{the unique integer in the interval } (x - 1, x]$$

$$\text{ceiling}(x) = \lceil x \rceil = \text{the unique integer in the interval } [x, x + 1)$$

For example, $\lfloor 4.1 \rfloor = 4$, $\lceil 4.1 \rceil = 5$, $\lfloor 4.0 \rfloor = 4$, $\lceil 4.0 \rceil = 4$.
Also, $\lfloor -4.1 \rfloor = -5$, $\lceil -4.1 \rceil = -4$, $\lfloor -4.0 \rfloor = -4$, $\lceil -4.0 \rceil = -4$.

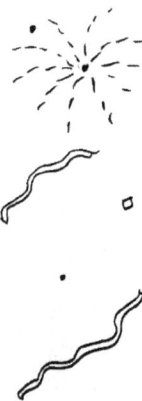

The grand finale for today will be the following theorem, which is perhaps the most significant fact about the integers.

Theorem *The Fundamental Theorem of Arithmetic*

Every integer n, with $n \geq 2$, can be uniquely written as a product of primes in ascending order.

I have used it already when I discussed the prime factorization of an integer.

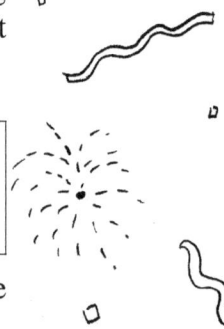

Definitions

The Rational Numbers The set of *rational numbers* is denoted by $\mathbb{Q}$, and is defined by

$$\mathbb{Q} = \left\{ \frac{p}{q} \;\middle|\; p \in \mathbb{Z}, q \in \mathbb{Z}, \text{ and } q \neq 0 \right\}$$

Divisible The integer a is *divisible* by the nonzero integer b if $a = bc$ for some integer c. We denote this by $b \mid a$ and also say that b *divides* a. We call b and c *divisors* of a.

Prime, Composite An integer, $p > 1$, is *prime* if its only divisors are 1 and p. An integer, $n > 1$, is *composite* if it is not prime. The number 1 is neither prime nor composite.

Even, Odd An integer, n, is *even* if there is an integer k such that $n = 2k$. An integer is *odd* if it is not even.

Greatest Common Divisor Let a and b be integers that are not both 0. The *greatest common divisor* (gcd) of a and b is a positive integer d such that

- $d \mid a$ and $d \mid b$.
- If c divides both a and b, then $c \mid d$.

The gcd of a and b is denoted by $\gcd(a, b)$.

Least Common Multiple The *least common multiple* (lcm) of two integers a and b is a nonnegative integer, m, such that

- $a \mid m$ and $b \mid m$.
- If both a and b divide c, then $m \mid c$.

The lcm of a and b is denoted by $\mathrm{lcm}(a, b)$.

Relatively Prime Positive integers a and b are *relatively prime* if $\gcd(a, b) = 1$.

Definitions

a* mod *m Let m be a positive integer. Then a mod m is the remainder when a is divided by m. The integer m is called the *modulus*.

***a ≡ b* (mod *m*)** We say that a is *congruent to* b, mod m, if m divides $a - b$. This is often expressed as: $a \equiv b \pmod{m}$ if and only if there is an integer k for which $a - b = km$.

Floor Function; Ceiling Function The *floor* and *ceiling* functions are defined for all real numbers x:

$$\text{floor}(x) = \lfloor x \rfloor = \text{the unique integer in the interval } (x - 1, x]$$

$$\text{ceiling}(x) = \lceil x \rceil = \text{the unique integer in the interval } [x, x + 1)$$

Theorems

The Quotient–Remainder Theorem Let a and b be integers with $b \neq 0$. Then there exist unique integers q and r such that $a = bq + r$ and $0 \leq r < |b|$.

gcd(a, b) = $as + bt$ Let a and b be integers such that at least one is not 0. Then there are integers, s and t, such that $\gcd(a, b) = as + bt$.

The Fundamental Theorem of Arithmetic Every integer n, with $n \geq 2$, can be uniquely written as a product of primes in ascending order.

Exercises

Solutions can be found at `http://www.mathcs.bethel.edu/~gossett/DMGN/`.

1. For each pair of integers, x and y, find the q and r (from the Quotient–Remainder theorem) that make $x = yq + r$. Express the results in a table. Also, determine whether $y \mid x$.

	x	y	q	r	$y \mid x$
(a)	295	38			
(b)	348	117			
(c)	1703	131			
(d)	4632	1442			

2. For each pair of integers, x and y, find the prime factorization of x and y and then find $\gcd(x, y)$. Express the results in a table.

	x	y	prime factorization of x	prime factorization of y	$\gcd(x, y)$
(a)	126	312			
(b)	315	1350			
(c)	360	10800			
(d)	297	490			

3. For each pair of integers, x and y, find the prime factorization of x and y and then find $\operatorname{lcm}(x, y)$. Express the results in a table.

	x	y	prime factorization of x	prime factorization of y	$\operatorname{lcm}(x, y)$
(a)	126	312			
(b)	315	1350			
(c)	360	10800			
(d)	297	490			

4. Find at least four integers that are congruent to 8, mod 13.

5. Is the following statement true or false?

   "The first five prime numbers are: 1, 2, 3, 5, 7."

6. For each pair of integers, x and y, find $x \bmod y$. Express the results in a table.

	x	y	$x \bmod y$
(a)	48	13	
(b)	45	66	
(c)	2064	444	
(d)	2090	38	

7. Let $\gamma(x) = \lceil x \rceil - \lfloor x \rfloor$. Show that

$$\gamma(x) = \begin{cases} 0 & x \in \mathbb{Z} \\ 1 & x \notin \mathbb{Z} \end{cases}$$

8. Consider the integers 37 and 530. By the Quotient–Remainder theorem, there exist *unique* integers, q and r, such that $37 = 530q + r$. However, note that $37 = 530 \cdot 0 + 37$ and $37 = 530 \cdot (-1) + 567$. Resolve the apparent contradiction.

9. Let $a, b, c, d, m \in \mathbb{Z}$ with $m > 0$. If $a \equiv b \pmod{m}$ and $c \equiv d \pmod{m}$, make a conjecture about the congruence of $a + c$ and $b + d$. Prove your conjecture.

10. Notice that $\gcd(11, 13) = 1$. Try to guess values for s and t such that $1 = 11s + 13t$.

Chapter 5

Hi Isolde. Did you solve the puzzle from last week?

Yes, I did. I am pretty proud of how I did it. Remember that Logan and I came up with the following diagram that gave an 8-person solution.

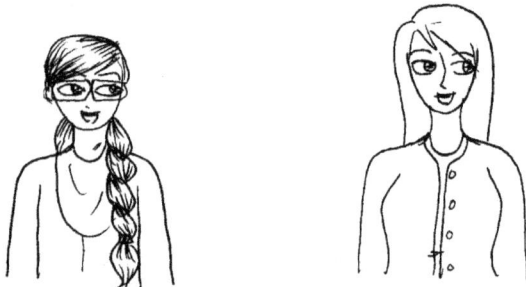

Solid single arrow means parent – child.
Dashed double arrow means spouse.

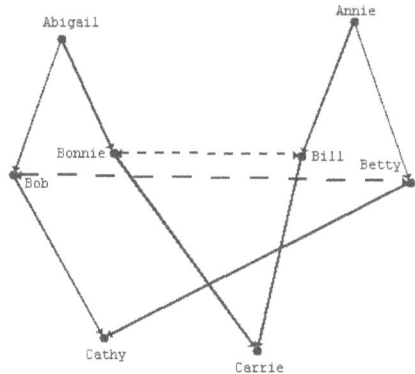

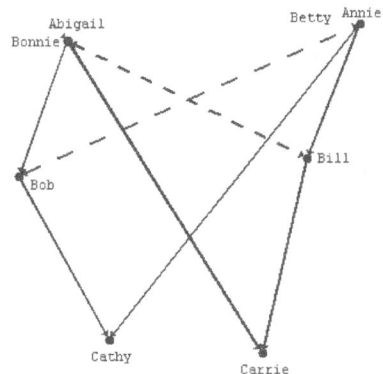

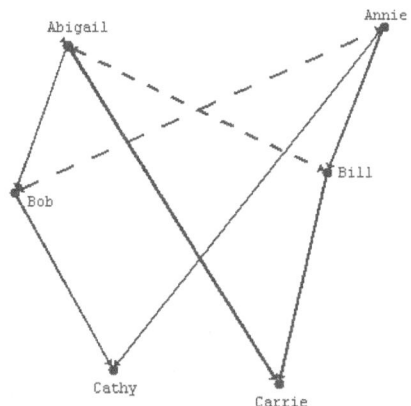

I needed to eliminate two people, and it needed to be done by merging. I needed to keep the two grandmothers and also the two unmarried granddaughters. I also needed to keep the two sons/husbands/fathers. So Bonnie and Betty seemed to be the people most likely to be eliminated. I still needed to keep two marriages, so Bonnie and Betty needed to merge with Abigail and Annie. The "no incest" rule implied that Abigail and Bonnie should merge and also Annie and Betty.

Solid single arrow means parent – child.
Dashed double arrow means spouse.

2 Grandmothers with their 2 Granddaughters
    Abigail–Cathy; Annie–Carrie
2 Husbands with their 2 Wives
    Bob–Annie; Bill–Abigail
2 Fathers with their 2 Daughters
    Bob–Cathy; Bill–Carrie
2 Mothers with their 2 Sons
    Abigail–Bob; Annie–Bill
2 Maidens with their 2 Mothers
    Cathy–Annie; Carrie–Abigail
2 Sisters with their 2 (Half)Brothers
    Carrie–Bob; Cathy–Bill

Solid single arrow means parent – child.
Dashed double arrow means spouse.

That is very good, Isolde!

Thank you, teacher. (^_^)

Now I have a puzzle for *you*, Lily. I have a basket with 10 apples. I also have 10 friends, who each wants an apple. So I give one apple to each friend. Afterward, there is still one apple in the basket. How is that possible?

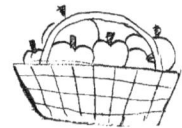

I suppose the unspoken puzzle rules don't allow a solution where one of the friends changes her mind and returns the apple.

No, it is not a trick answer, just one where you need to be a bit creative in your thinking.

Oh! I get it! You can hand the last apple *still sitting in the basket* to the last friend.

Good job! Ok, before we start today's topic, I thought I would share another study tip. Here is another quote from *Make it stick: the Science of Successful Learning*, by Brown, Rodediger, McDaniel.

Cramming. a form of massed practice, has been likened to binge-and-purge eating. A lot goes in, but most of it comes right back out in short order. The simple act of spacing out study and practice in installments and allowing time to elapse between them makes both the learning and the memory stronger, in effect building habit strength.

I have seen many of my university classmates resort to cramming for exams. It is not very effective. First, as the quote mentions, there is no retention. Even worse, that style of "learning" does a poor job of building understanding. There is no time for your brain to work on making the proper connections (something that happens during sleep).

A better approach is to study new material and then review it later. This review and rehearsal builds stronger connections in the brain and also helps to move the connections to long-term memory.

Then I promise to never cram for exams!

Don't miss the other part of what I said—study new material more than once. Often school schedules don't give you much time to review, but if you are deliberate about it, you can make time. If you do, you should see your grades go up.

Actually, my grades are already pretty high.

That is true, but the real goal is to learn more thoroughly. And we can all improve on that.

Let's start today's session.

Proof

Theorem *The Infinitude of the Primes.*

There is an infinite number of distinct primes.

Proof:

Suppose instead that there are only a finite number of primes. Denote them as $\{p_1, p_2, p_3, \ldots, p_n\}$, where n is a positive integer. Consider the positive integer $q = (p_1 p_2 \cdots p_n) + 1$. Since $q \notin \{p_1, p_2, p_3, \ldots, p_n\}$, q must be composite. What are the prime factors of q? (See the Fundamental Theorem of Arithmetic on page 85.)

If $p_k \mid q$, then there is an integer m with $q = mp_k$ so $mp_k = (p_1 p_2 \cdots p_n) + 1$. Thus $(p_1 p_2 \cdots p_n) - mp_k = -1$ and so $p_k(p_1 p_2 \cdots p_{k-1} p_{k+1} \cdots p_n - m) = -1$. This implies p_k is a factor of -1, which is impossible.

Since none of the primes p_k divide q and $\{p_1, p_2, p_3, \ldots, p_n\}$ are all the primes that exist, q cannot be composite. It must be the case that q is a new prime. This contradicts the assumption that there is only a finite number of primes and they were all listed in $\{p_1, p_2, p_3, \ldots, p_n\}$.

Therefore, there must be an infinite number of primes.

$\square$

Note:

The proof led to the conclusion that $(p_1 p_2 \cdots p_n) + 1$ is a prime only because we assumed that there were a finite number of primes. The conclusion is not valid in general. For example, 2, 5, and 11 are all primes but $(2 \cdot 5 \cdot 11) + 1 = 111 = 3 \cdot 37$ is composite.

Hi Isolde. What is the topic for today?

I thought I would outline some of the basic ideas associated with reading and creating proofs.

Before I go into specific proof strategies, I need to describe the context in which we do proofs. I have previously mentioned that sets, logic, and the axiomatic system are the basis for modern mathematics. The new piece is the axiomatic system.

SETS LOGIC
THE AXIOMATIC SYSTEM

The key insight of the axiomatic system is to realize that all ideas need to be understood in terms of other ideas. So there is potentially an infinite regress of concepts if we keep trying to push back to the start.

In the axiomatic system, we start with some undefined terms. Way back around 300 BCE, Euclid wrote about geometry. He started with some terms that he never defined. For example, *point* and *line*. We have intuitive ideas about what these should mean, but they are just taken as objects whose behavior and relationships we specify.

Undefined Terms
Axioms

We specify these relationships using *axioms*. Axioms are statements whose validity we **assume** to be true. We are only interested in situations where the axioms are true. We try to choose axioms that are simple, intuitive, and non-redundant (*independent*). For example, Euclid chose 5 axioms for his geometry.

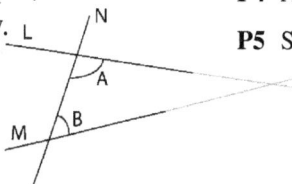

P1 A straight line (segment) can be drawn from any point to any other point.

P2 A line segment can be extended continuously to form a line.

P3 A circle can be created having any center and radius.

P4 All right angles are equal to one another.

P5 Suppose line N crosses lines L and M, and the angles A and B (both on the same side of N) add to less than two right angles. Then lines L and M will eventually meet, on the side of N where A and B lie.

The next components of the axiomatic system are definitions. A *definition* uses a word or short phrase to represent a more complex concept. For example, in geometry we define a *circle* to be the collection of points in the plane that are equidistant from a given point, which is called the *center of the circle*.

Everything else in the axiomatic system are things that require proof.

Things that don't need proof
Undefined terms
Axioms
Definitions
Things that require proof

I will mention 4 things that require proof. The definitions I use are not universal, but are very common. *Theorems* and *Propositions* are major assertions about the objects in the axiomatic system. I use those terms interchangeably, but will mostly choose the word *theorem*. *Corollaries* are assertions that are simple or immediate consequences of a theorem's truth. *Lemmas* are assertions that are usually not important on their own but serve to simplify the proof of some more important theorem.

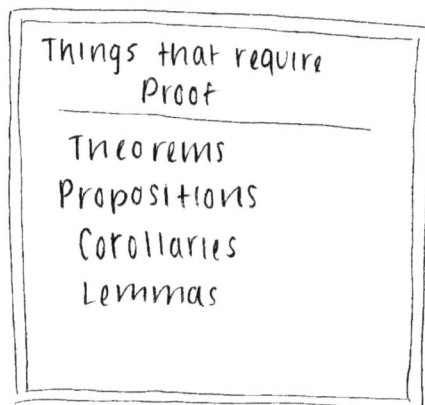

Things that require Proof
Theorems
Propositions
Corollaries
Lemmas

Informally, a **proof** is a demonstration of the validity of some precise mathematical statement.
The demonstration should contain sufficient detail to convince the intended audience of its validity.

More formally, a *mathematical proof* of the statement S is a sequence of logically valid statements that connect axioms, definitions, and other already validated statements into a demonstration of the correctness of S. The rules of logic and the axioms are agreed on ahead of time. At a minimum, the axioms should be independent and consistent. The amount of detail presented should be appropriate for the intended audience.

That is sufficient background to start looking at proof strategies. There are many proof strategies available. I will discuss three of the most common: *direct proof, indirect proof,* and *proof by contradiction.*

Indirect proof

Mary, go ask Sue why she is angry with me.

Proof by contradiction

Don't talk back to *me* young lady!

The proof strategies are based upon some logical equivalences (page 36) and rules of inference. Let A and B be two propositions (page 34). Then B may be *inferred* from A, denoted by $A \Rightarrow B$, if $A \rightarrow B$ is a tautology (page 37). The meta-statement $A \Rightarrow B$ is called a *rule of inference.*

The point is that if $A \rightarrow B$ is always true **and if** A is known to be true, then B must be true (look at the truth table for implication). The truth of A eliminates the two bottom rows. The truth of $A \rightarrow B$ eliminates the second row. That means we must be in the first row, and B is true in that row.

A	B	$A \rightarrow B$
T	T	T
T	F	F
F	T	T
F	F	T

This idea is captured in the inference rule *Modus Ponens*: $[P \wedge (P \rightarrow Q)] \Rightarrow Q$

Another useful rule of inference is the *Law of Hypothetical Syllogism*:
$[(P \rightarrow Q) \wedge (Q \rightarrow R)] \Rightarrow (P \rightarrow R)$

Putting these together leads to the strategy of *direct proof*: start with the assumed truth of P and show that $P \rightarrow Q$ is true. Then (modus ponens) we conclude that Q is true.

Now prove that $Q \rightarrow R$ is true and conclude that R is true. we can easily extend this to have more intermediate propositions (instead of only Q).

Here is an example of a direct proof.

Let a and b be real numbers. Then

$$(a + b)^2 = a^2 + 2ab + b^2$$

The proof uses many of the familiar properties (actually axioms) from the system of real numbers.

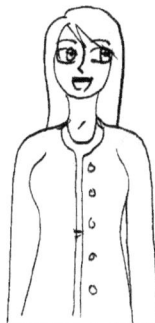

DIRECT
PROOF

$$
\begin{aligned}
(a + b)^2 &= (a + b)(a + b) && \text{definition of "squared"} \\
&= (a + b)a + (a + b)b && \text{distributivity} \\
&= a(a + b) + b(a + b) && \text{commutativity (twice)} \\
&= [aa + ab] + [ba + bb] && \text{distributivity (twice)} \\
&= [aa + ab] + [ab + bb] && \text{commutativity} \\
&= ([aa + ab] + ab) + bb && \text{associativity} \\
&= (aa + [ab + ab]) + bb && \text{associativity} \\
&= (aa + [ab \cdot 1 + ab \cdot 1]) + bb && \text{identity (twice)} \\
&= (aa + ab \cdot [1 + 1]) + bb && \text{distributivity} \\
&= (aa + 2ab) + bb && \text{addition and commutativity} \\
&= (a^2 + 2ab) + b^2 && \text{definition of "squared" (twice)} \\
&= a^2 + 2ab + b^2 && \text{associativity}
\end{aligned}
$$

That looks a lot like the kind of proofs we did with sets and logic.

Yes, it does. Here is another theorem, whose direct proof is formatted differently.

Let a and b be real numbers, with $a \geq 0$ and $b \geq 0$. Then $\frac{a+b}{2} \geq \sqrt{ab}$, with equality if and only if $a = b$.

Any square is nonnegative, so $(a - b)^2 \geq 0$. That is, $a^2 - 2ab + b^2 \geq 0$.

Since both a and b are nonnegative, $4ab \geq 0$. Thus

$$(a + b)^2 = a^2 + 2ab + b^2 = (a^2 - 2ab + b^2) + 4ab \geq 4ab \geq 0$$

So $(a + b)^2 \geq 4ab \geq 0$. Since both sides of $(a + b)^2 \geq 4ab$ are nonnegative, taking square roots does not change the direction of the inequality. Thus, $(a + b) \geq 2\sqrt{ab}$ and so $\frac{a+b}{2} \geq \sqrt{ab}$. If $a = b$ it is clear that $\frac{a+b}{2} = \sqrt{ab}$. If $\frac{a+b}{2} = \sqrt{ab}$, square both sides to conclude that $(a - b)^2 = 0$ so $a = b$. $\square$

(The symbol $\square$ indicates the end of the proof.)

67

The next proof strategy is *indirect proof*. It is based on the logical equivalence of an implication and its contrapositive (page 38).

$$[P \rightarrow Q] \Leftrightarrow [(\neg Q) \rightarrow (\neg P)]$$

The key idea is to prove the contrapositive (the implication $\neg Q \rightarrow \neg P$). Once that is done, the logical equivalence ensures that the original implication ($P \rightarrow Q$) must also be true.

I am not sure why that makes things any easier.

Sometimes the propositions, P and Q, are a bit slippery but the propositions $\neg P$ and $\neg Q$ are more concrete.

Recall that a *rational number* is one that can be expressed in the form $\frac{p}{q}$, where both p and q are integers and $q \neq 0$. An *irrational number* is a real number that is not rational.

Consider the claim

Let x be a real number. If x is irrational, then $2x$ is irrational.

How would you prove this claim?

I suppose I would start by assuming that x is irrational and work from there. But what does it mean for x to be irrational? I only know I *can't* write it in the form $\frac{a}{b}$ where a and b are integers. So how can I write it?

That is indeed the central issue: there is no concrete way to specify x other than by what it is *not*. Fortunately, indirect proof steps in to save the day. The contrapositive reads:

Let x be a real number. If $2x$ is rational, then x is rational.

(Notice that "x is not irrational" is the same thing as "x is rational.") But the claim "$2x$ is rational" is something that can be expressed quite concretely: there exist integers, a and b, with $b \neq 0$ such that $2x = \frac{a}{b}$. The full proof is:

If $2x$ is rational, then there exist integers, a and b, with $b \neq 0$ such that $2x = \frac{a}{b}$.

But then $x = \frac{a}{2b}$. Since both 2 and b are integers, $2b$ is also an integer. And since the product of two nonzero integers is a nonzero integer, $\frac{a}{2b}$ is a rational number. □

Since the contrapositive is true, so is the original implication.

Let's consider a new assertion.

Let n be a positive integer. If $2^n - 1$ is prime, then n is prime.

Which is easier, a direct proof or an indirect proof?

Lily thinks a while.

I am not sure either of them is easier. If I try a direct proof, I am stuck right away, since I don't know how to move from $2^n - 1$ being prime to the conclusion that n is prime. On the other hand, if I start by assuming that n is not a prime then I can conclude that there are positive integers a and b with $n = ab$. But I don't know how to show that $2^{ab} - 1$ can be written as a composite number.

Is $2^{ab} - 1 = xy$ where x and y are integers with $x > 1$ and $y > 1$?

Actually, you have made a very good start to show that an indirect proof is the better approach. Writing $n = ab$ is something concrete, so there is hope for progress. You just got stuck on how to show that $2^{ab} - 1$ is composite. Normally I would have you make a table of n and $2^n - 1$ for many values of $n > 1$ and see if $2^n - 1$ factors when n is composite.

However, that will sidetrack us from the main point right now, so I will provide that part of the proof. The key idea is that if $a \leq b$, then $2^{ab} - 1 = xy$, where $x = (2^a - 1)$ and $y = \left(2^{ab-a} + 2^{ab-2a} + \cdots + 2^{2a} + 2^a + 1\right)$. (This is easy to see if it's written down—just multiply the two expressions and simplify.)

Since $a > 1$ and $b > 1$, it is also true that $x > 1$ and $y > 1$, so $2^{ab} - 1$ is not prime.

That makes sense, but I don't think I could have finished on my own.

Given enough time, I think you might have. You made a good start. You just needed the time to experiment and look for a pattern.

The final proof strategy we will consider is *proof by contradiction*. We start with $P \to Q$ and then immediately assume Q is false. It is like a 2-year-old's response: "no!"

"no!"

There are several logical equivalences that support this strategy. All start by assuming we are in row 2 of the implication truth table. That is, assume P is true but Q is false. Eventually, derive a contradiction: such as R and its opposite are both true $R \land \neg R$, or P is actually false, or Q is actually true.

P	Q	$P \to Q$
T	T	T
T	F	F
F	T	T
F	F	T

Assume we are here $\longrightarrow$ (row 2)

$$[P \to Q] \Leftrightarrow [(P \land (\neg Q)) \to (R \land (\neg R))]$$
$$[P \to Q] \Leftrightarrow [(P \land (\neg Q)) \to (\neg P)]$$
$$[P \to Q] \Leftrightarrow [(P \land (\neg Q)) \to Q]$$

What went wrong? It started with the assumption $P \land \neg Q$, so we must conclude that row 2 of the implication truth table cannot be valid for this specific implication.

I think I get it, but an example would be helpful.

Here is a proof by contradiction.

Let x be a real number. If x is irrational and x^2 is rational, then x^3 is irrational.

I will start the proof by assuming that x is irrational and x^2 is rational, but x^3 is rational.

Notice that P is in the form $P = A \land B$ where A is "x is irrational" and B is "x^2 is rational." Q is the assertion "x^3 is rational."

Consequently, there exist integers a, b, c, d with $b, d \neq 0$ such that $x^2 = \frac{a}{b}$ and $x^3 = \frac{c}{d}$. But then (a can't be 0 or else x would be 0)

$$x = \frac{x^3}{x^2} = \frac{c \cdot b}{d \cdot a} = \frac{n}{m} \text{ with } m \neq 0$$

But this means that x is a rational number, which contradicts the original assumption that x is irrational.

So you used the logical equivalence
$$[P \to Q] \Leftrightarrow [(P \land (\neg Q)) \to (\neg P)]$$
where $\neg P = \neg(A \land B)$
$= \neg A \lor \neg B$. In this case, $\neg A$ was the contradiction.

Here is a more significant proof by contradiction. The earliest recorded version we have is in Euclid's *Elements*, which dates back to around 300 BC. Euclid probably wasn't the original creator of the proof.

Theorem $\sqrt{2}$ *is irrational*

70

The real number $\sqrt{2}$ is irrational.

Proof:

Assume instead that $\sqrt{2}$ *is* a rational number. Therefore, there are integers p and $q \neq 0$ such that $\sqrt{2} = \frac{p}{q}$. We will assume that p and q have no common factors. (If there were any, we could factor them out and cancel, leaving us with a new p and q.) Since q is not 0, we can multiply the equation by q, leaving $\sqrt{2}q = p$. Squaring both sides produces $2q^2 = p^2$. Since the left-hand side has a factor of 2, the right-hand side does also. This means that 2 is a factor of p.

Hence, $p = 2r$, for some integer r. The equation $2q^2 = p^2$ can be rewritten as $2q^2 = 4r^2$. Dividing by 2 produces $q^2 = 2r^2$, from which we conclude (in a now familiar way) that 2 is a factor of q. This is the contradiction we need: 2 is a factor of both p and of q, but p and q have no common factors.

The trouble began with the assumption that $\sqrt{2}$ is rational. Hence, $\sqrt{2}$ must be irrational.

□

I can't talk to you- you are irrational!

So which logical equivalence is behind this proof? I am not even sure how to state the original implication.

That is a good question. We might state the implication as:

If a real number is $\sqrt{2}$, then that number is irrational.

The logical equivalence would be
$$[P \rightarrow Q] \Leftrightarrow [(P \wedge (\neg Q)) \rightarrow (R \wedge (\neg R))]$$
where R is that assertion that p and q have no common factors.

We are out of time, but that is everything I planned to discuss today. Here are some exercises to try for homework.

Ok. This will take some time to process. See you next week.

Axiomatic Mathematics

The axiomatic method is a technique of deduction from prior concepts. In order for such a system to get started, some prior concepts must exist that do not need to be based on still earlier concepts. Mathematicians start with undefined terms. *Undefined terms* are usually ideas that have enough intuitive appeal that we may safely use them as a starting place.

In addition, a set of properties that the undefined terms satisfy is needed. Such properties, which are *assumed* to be true, are called *axioms*. The valid deductions that exist in the system are directly dependent on the set of axioms chosen.

As the axioms lead to additional concepts, it is convenient to give some of them names, especially if the concept is nontrivial. A *definition* is the means for binding a concept, a name for the concept, and a set of associated properties that describe the concept. A definition is a form of aliasing. The concept (name) is equated with the properties. As such, we may substitute the properties for the concept whenever the concept is known to occur, or to conclude that the concept occurs if the properties are shown to hold.

Any statement that is not an axiom or definition needs to be proved. Important statements that have been proved are called *theorems* or *propositions*. Sometimes the proof of a theorem is quite long. It is often useful to modularize long proofs by introducing subtheorems called *lemmas*. A lemma is usually considered to be a mini-theorem whose main purpose for existing is to help prove part of a more important theorem or proposition. A *corollary* is a statement whose truth is an immediate consequence of some other theorem or proposition.

Proof

A *mathematical proof* of the statement S is a sequence of logically valid statements that connect axioms, definitions, and other already validated statements into a demonstration of the correctness of S. The rules of logic and the axioms are agreed on ahead of time. At a minimum, the axioms should be independent and consistent. The amount of detail presented should be appropriate for the intended audience.

Logical Equivalences and
Rules of Inference for Theorems

Modus Ponens

$[P \wedge (P \rightarrow Q)] \Rightarrow Q$

Law of Hypothetical Syllogism

$[(P \rightarrow Q) \wedge (Q \rightarrow R)] \Rightarrow (P \rightarrow R)$

Contrapositive (Indirect Proof)

$[P \rightarrow Q] \Leftrightarrow [(\neg Q) \rightarrow (\neg P)]$

Proof by Contradiction

$[P \rightarrow Q] \Leftrightarrow [(P \wedge (\neg Q)) \rightarrow (R \wedge (\neg R))]$

$[P \rightarrow Q] \Leftrightarrow [(P \wedge (\neg Q)) \rightarrow (\neg P)]$

$[P \rightarrow Q] \Leftrightarrow [(P \wedge (\neg Q)) \rightarrow Q]$

Proof Strategies

Three common strategies to prove the implication $P \rightarrow Q$.

Direct Proof Assume that P is true and use a sequence of valid assertions to arrive at the conclusion that Q is true.

Indirect Proof Assume that Q is false and use a sequence of valid assertions to arrive at the conclusion that P is also false, thus proving that the contrapositive $\neg Q \rightarrow \neg P$ is true. But the original implication and the contrapositive are logically equivalent, so the original implication must also be true.

Proof By Contradiction Assume that P is true but Q is false. Use a sequence of valid assertions to arrive at a contradiction. Conclude that it is not possible for P to be true and Q to be false at the same time. So one of the other three rows of the implication truth table must be valid. In all three cases, the implication is true.

Exercises

Solutions can be found at http://www.mathcs.bethel.edu/~gossett/ DMGN/.

Definition An integer, n, is *even* if and only if there is an integer, k, such that $n = 2k$.

Definition An integer, n, is *odd* if and only if it is not even.

Notes: If n is odd, then there is an integer k such that $n = 2k + 1$. The integer 0 is even because $0 = 2 \cdot 0$.

1. Suppose the integer n is even. Use a direct proof to show that n^2 is also even.

2. Suppose the integer n is odd. Use an indirect proof to show that n^2 is also odd. (Hint: If 2 is a factor of n^2, then it is also a factor of n.)

3. Use a direct proof to show that the sum of an even and an odd integer is odd.

4. Let n be integer. Use an indirect proof to show that if $n^3 + 1$ is odd, then n is even.

5. Use a proof by contradiction to show that every prime that is greater than 2 is an odd number.

6. Let A and B be sets. Use a proof by contradiction to show that $(A \cap B) - B = \emptyset$.

7. Use any of the strategies to prove that the sum of a rational number and an irrational number is irrational.

8. Let a and b be nonzero integers. If a is even and b is odd, show that when dividing b by a, the remainder must be odd. (Hint: Write the division in the form $b = aq + r$.)

9. Let A and B be sets. Prove that $(A - B) - A = \emptyset$.

10. Can the product of two irrationals be a rational? If not, provide a proof that this cannot happen. Otherwise give a *counterexample* to the assertion "the product of two irrationals is irrational."

Chapter 6

Isolde, I have a question before we start. Logan and I are second-generation Americans—our parents came here from Taiwan for graduate school and later became naturalized citizens. What is your background? With a name like Isolde Gallagher, I would guess Irish.

Your guess is correct. My earliest ancestors who came to America arrived in 1848, right in the middle of the Great Famine, when the potato crops failed. The famine was a result of a crop blight and some pretty callous political decisions and mass evictions by greedy landlords. They came across the ocean on what were called coffin ships because so many of the passengers died during the crossings.

That must have been a terrible time. I know that there were many famines in China over the centuries. The most recent was during the 20th century.

People tend to remember events of that nature and tell their children. My family still talk about it 5 or 6 generations after it happened.

So, Isolde, do you believe in leprechauns? (^_^)

Of course not. However, I was always fascinated by another kind of being from Irish and Scottish mythology: the selkies.
I actually have a nice puzzle that involves selkies. Let me dig it out of my backpack.

That sounds fun. Maybe next week I can bring a Chinese puzzle for you to solve.

On a small island off the coast of Ireland is a community with some odd characteristics. The people living on the island are a mixture of humans and selkies. Selkies live as seals in the sea but can shed their skin and live as humans on land. On land, it is impossible to tell selkies from humans by sight. The inhabitants of the island also differ in the following manner. Some always tell the truth, and some always lie. Telling the truth or lying is independent of whether the inhabitant is human or selkie. Suppose you are walking on the island and meet four of the inhabitants. You learn that their names are Aiden, Brigid, Conall, and Dierdre. You ask them two questions and each inhabitant replies to the questions.

Are any of you liars?

> **Aiden:** Exactly one of us is lying.
>
> **Brigid:** Exactly two of us are lying.
>
> **Conall:** Exactly three of us are lying.
>
> **Dierdre:** Exactly four of us are lying.

Are any of you a selkie?

> **Aiden:** Conall is a selkie.
>
> **Brigid:** Aiden is a selkie.
>
> **Conall:** Only one of us is a selkie.
>
> **Dierdre:** Brigid is not a selkie.

Which of the inhabitants are liars and which are truthful? Can you identify any of them as human or selkie?

We interrupt your regularly scheduled reading to give you time to think about this logic problem. Perhaps a picture of Lily feeding seals at the zoo will spur your imagination.

77

I have the solution. At first I just started randomly guessing who told the truth and who lied, but I gave up on that and decided to be systematic. There are 16 possible combinations of who is **Honest** and who is **Lying**. I built a table and then looked at the first question and decided which of the H/L combinations were consistent with the known answers.

For example, in the row that starts HLHL, A tells the truth but there are 2 liars, so that is not consistent with his answer. I recorded an N to indicate the inconsistency. Since B is a liar and B's answer to question 1 is consistent with two liars, I also record an N (since B's answer is true, but B does not tell the truth).

I eventually found that only the row LLHL was consistent for all four answers. Once I knew who was lying and who was honest, it was easy to determine that Brigid is the only selkie.

Veracity				Q1				Veracity				Q1			
A	B	C	D	A	B	C	D	A	B	C	D	A	B	C	D
H	H	H	H	N	N	N	N	L	H	H	H	N	N	N	N
H	H	H	L	Y	N	N	Y	L	H	H	L	Y	Y	N	N
H	H	L	H	Y	N	Y	N	L	H	L	H	Y	Y	Y	N
H	H	L	L	N	Y	Y	Y	L	H	L	L	Y	N	N	N
H	L	H	H	Y	Y	N	N	L	L	H	H	Y	N	N	N
H	L	H	L	N	N	N	Y	**L**	**L**	**H**	**L**	**Y**	**Y**	**Y**	**Y**
H	L	L	H	N	N	Y	N	L	L	L	H	Y	Y	N	N
H	L	L	L	N	Y	N	Y	L	L	L	L	Y	Y	Y	N

That is correct. However, there is a way to avoid producing the entire table. Notice that at most one of the four answers to the first question can be true. Start with Dierdre. Her statement can't be true, because she would be one of the four liars. So we know she is lying and eliminate the LLLL row.
If Conall is telling the truth, then we are in the consistent LLHL row. Otherwise, Conall is lying so we can ignore all rows with three Ls.
If Brigid is honest, then we only need to investigate (and eliminate) the rows HHLL, LHHL, LHLH (B must be H). Otherwise, we can skip rows with exactly two Ls.
If Aiden is telling the truth, we just need to investigate HLLL (and eliminate it). Otherwise, none of the rows with just one L can be valid.
Only 5 of the rows need be carefully examined: LLHL is the solution.

Mathematical Induction

Theorem *Partial Sum of a Geometric Progression*
Let $r \in \mathbb{R}$ with $r \neq 0$ and $r \neq 1$. Then

$$\sum_{i=0}^{n} r^i = \frac{1 - r^{n+1}}{1 - r} = \frac{r^{n+1} - 1}{r - 1}$$

Base Step
When $n = 0$, $\sum_{i=0}^{0} r^i = r^0 = 1$ and $\frac{1 - r^{0+1}}{1-r} = 1$ since $r \neq 1$.

Inductive Step
Assume that $\sum_{i=0}^{n} r^i = \frac{1 - r^{n+1}}{1-r}$ is true for some $n \geq 0$. We want to show that

$$\sum_{i=0}^{n+1} r^i = \frac{1 - r^{n+2}}{1 - r} \quad \text{is also true.}$$

$$\sum_{i=0}^{n+1} r^i = \left(\sum_{i=0}^{n} r^i \right) + r^{n+1}$$

$$= \left(\frac{1 - r^{n+1}}{1 - r} \right) + r^{n+1} \quad \text{by the inductive hypothesis}$$

$$= \left(\frac{(1 - r^{n+1}) + (1 - r)r^{n+1}}{1 - r} \right)$$

$$= \frac{1 - r^{n+2}}{1 - r}$$

Conclusion
The claim is true when $n = 0$ and whenever it is true for some $n \geq 0$ it is also true for $n + 1$. The Mathematical Induction theorem implies that the claim is true for all $n \geq 0$. $\square$

Hello, Lily. I hope you came ready to concentrate. Today's topic is induction.

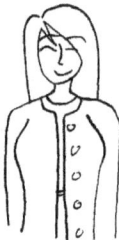

Like being inducted into the Baseball Hall of Fame or the military?

No. We will be looking at *proof by induction*. It is a proof strategy that is frequently used in discrete mathematics. We can use it when the theorem is making a claim like "for all natural numbers n, $n^2 \geq n$."

The feature on which to focus is the "for all natural numbers" part. It need not be natural numbers. For example, the claim might be stated about all integers $n \geq 1$ or all integers $n > 4$. The common characteristic is that the claim is asserted for all integers greater than or equal to some initial value.

The proof first shows that the claim is true for the initial value of n. It then shows that whenever the claim is true for some integer k (that is greater than or equal to the initial value), then it is also true for $k + 1$. It is very much like knocking down a chain of dominoes. If we know that the first domino falls down and we know that each domino knocks down the next domino, then we are sure that all the dominos will fall.

I actually like knocking down domino chains. Is an induction proof that easy? I suspect that there is more to it than your analogy implies.

Yes, there is a bit more to it. There are two issues you need to deal with in an induction proof. The first is to properly follow the form of the proof. I will discuss that after the induction theorem is presented. The second issue shows up in the middle of the proof and usually involves doing some algebra that is totally unrelated to induction. Once you learn the form, the second issue is where students typically get stuck.

$$(n + 1)! = (n + 1) \cdot n!$$
$$\leq (n + 1) \cdot n^n$$

How do I get to the end from here?

$$\leq (n + 1)^{n+1}$$

Here is the theorem that validates the strategy.

Theorem *Mathematical Induction*

If $\{P(n)\}$ is a set of statements such that

1. $P(1)$ is true   and

2. $P(i) \rightarrow P(i + 1)$ for $i \geq 1$

then $P(n)$ is true for all positive integers n. This can be stated more succinctly as

$$[P(1) \wedge (\forall i, P(i) \rightarrow P(i + 1))] \rightarrow [\forall n \geq 1, P(n)]$$

The symbol $\forall$ means *for all*. A proof by induction is structured to show that the two parts of the hypothesis are satisfied. Once that is done, the theorem guarantees that $P(k)$ is true for all $k \geq 1$.

Can you show an example of a proof that uses induction?

Of course. (^_^)

Theorem *Sum of the first n positive integers*

Let $n \geq 1$. Then

$$\sum_{i=1}^{n} i = \frac{n(n + 1)}{2}$$

I will break the proof into three phases: a base step, an inductive step, and a conclusion.

Base Step

When $n = 1$ we see that $\sum_{i=1}^{1} i = 1 = \frac{1(1+1)}{2}$, so the theorem is true in this case. (A brief summation notation review is on page 230.)

Inductive Step

Assume that there is some $n \geq 1$ for which $\sum_{i=1}^{n} i = \frac{n(n+1)}{2}$ is true. *It is important to explicitly state this assumption and also that we use the word "some" and not the word "all."*

The goal is to now show that $\sum_{i=1}^{n+1} i = \frac{(n+1)(n+2)}{2}$ is also true. (I replaced every "n" with "n + 1.") This will show that the implication in hypothesis 2 of the mathematical induction theorem is true. I will start with the left-hand side of the new equation and derive the right-hand side.

A common way to start $\longrightarrow$
$$\sum_{i=1}^{n+1} i = \sum_{i=1}^{n} i + (n + 1) \qquad \text{peel off the final summand}$$

We assumed that $\sum_{i=1}^{n} i = \frac{n(n+1)}{2}$ $\longrightarrow$

Always label where you use the ind. hyp.
$$= \frac{n(n + 1)}{2} + (n + 1) \qquad \text{by the inductive hypothesis}$$

$$= \frac{(n + 1)(n + 2)}{2} \qquad \text{some simple algebra}$$

81

Conclusion

I have shown that the claim is true when $n = 1$ and also that if the claim is true for some n, then it is also true for $n + 1$. The Mathematical Induction theorem implies that the claim is true for *all* $n \geq 1$.

I have defined $P(n)$ to be the assertion $\sum_{i=1}^{n} i = \frac{n(n+1)}{2}$. I first showed that $P(1)$ is true, and then showed that $P(n) \rightarrow P(n + 1)$ is true for $n \geq 1$. I then conclude that $P(n)$ is true for all $n \geq 1$.

That kind of makes sense, but I have a question. I see how, in the inductive step, we assume that $P(n)$ is true and then show that $P(n + 1)$ must also be true. But why can we assume that $P(n)$ is true? Isn't that what we need to prove?

That is a good question. What we are doing in the inductive step is proving that the *implication* $P(n) \rightarrow P(n + 1)$ is true. In order to do that, we assume that the hypothesis, $P(n)$, is true (if it is false, the implication is automatically true—see the truth table for implication on page 37). We then show that the conclusion, $P(n+1)$, is true, ensuring that the implication is true.

Once the base and inductive steps are done, we can use modus ponens (page 66) to conclude that $P(2)$ is true:
$$P(1) \wedge \big(P(1) \rightarrow P(2)\big) \rightarrow P(2).$$

But then we can repeat the process to conclude that $P(3)$ is true:
$$P(2) \wedge \big(P(2) \rightarrow P(3)\big) \rightarrow P(3).$$
This process continues forever (the domino effect), so $P(n)$ must be true for all $n \geq 1$.

So we are not really assuming that $P(n)$ is true. Instead we say "*if* $P(n)$ is true, then $P(n + 1)$ must also be true." Is it possible to prove $P(n) \rightarrow P(n + 1)$ but not prove $P(1)$ is true? Wouldn't the whole system fail in that case?

Yes, here is such an example. Consider the claim $\ln(a^n) = (n + 1)\ln(a)$ for some fixed real number $a > 1$. So $P(n)$ is the claim: $\ln(a^n) = (n+1)\ln(a)$. I can easily show that $P(n) \rightarrow P(n + 1)$:

$$\ln\left(a^{n+1}\right) = \ln\left(a^n \cdot a\right) = \ln\left(a^n\right) + \ln(a).$$

By the inductive hypothesis this is equal to
$(n + 1)\ln(a) + \ln(a) = (n + 2)\ln(a).$
Thus $\ln\left(a^{n+1}\right) = (n + 2)\ln(a)$ and so $P(n + 1)$ is true.

But of course, $\ln(a^n) = n\ln(a)$ by the logarithm properties (page 229), so the claim is false.

It is time to see how you do on an induction proof, Lily. Here is your theorem:

Theorem

Let $n \in \mathbb{Z}$.
If $n > 1$ then $n^2 > n$.

Here is my proof.

Base Step

If $n = 2$ then $2^2 = 4 > 2$, so the claim is true.

Inductive Step

Assume that there is some $n > 1$ for which $n^2 > n$. Then

$$(n + 1)^2 > n + 1$$

so the inductive step is complete.

No, that doesn't feel right. I want to start with $(n + 1)^2$ and end up with $n + 1$, but I am not sure how to make the proof of the inequality anything more than "because I know it is true."

One problem I see is that you have not used the inductive hypothesis (the assumption that $n^2 > n$). You need to manipulate $(n + 1)^2$ to look like something involving n^2 (you are starting with the left-hand side of $P(n + 1)$ so you look for the left-hand side of $P(n)$).

Well, $(n + 1)^2 = n^2 + 2n + 1$ and that involves n^2. I suppose I could replace the n^2 with n and switch to a > sign. That would give me $(n + 1)^2 > 3n + 1$. Not quite what I want. Any hints?

You already have an inequality. Why not make the right-hand side *even smaller*.

Oh! I see. Here is my *new! improved!* inductive step.

Inductive Step

Assume that there is some $n > 1$ for which $n^2 > n$. Then

$$
\begin{aligned}
(n + 1)^2 &= n^2 + 2n + 1 && \text{algebra} \\
&> n + (2n + 1) && \text{by the inductive hypothesis} \\
&= 3n + 1 \\
&> n + 1 && 3n > n \text{ if } n > 0
\end{aligned}
$$

so the inductive step is complete.

Conclusion

The claim is true when $n = 2$ and whenever it is true for some $n \geq 2$ it is also true for $n + 1$. The Mathematical Induction theorem implies that it is true for all $n \geq 2$.

83

I have one more example before I move on to another induction theorem. I want to use mathematical induction to prove the following theorem.

Theorem

For integers n with $n \geq 1$

$$\sum_{i=1}^{n} i^2 = \frac{n(n+1)(2n+1)}{6}$$

Should I memorize this?

It is helpful in this example to look at the shape of $P(n+1)$ before starting the proof. The right-hand side of $P(n)$ contains $n(n+1)(2n+1)$ as its numerator. The numerator for that part of $P(n+1)$ would then be
$$(n+1)\big((n+1)+1\big)\big(2(n+1)+1\big) = (n+1)(n+2)(2n+3).$$

Base Step

When $n = 1$
$$\sum_{i=1}^{1} i^2 = 1^2 = 1 = \frac{1(1+1)(2 \cdot 1 + 1)}{6}$$

Inductive Step

Assume that $\sum_{i-1}^{n} i^2 = \frac{n(n+1)(2n+1)}{6}$ is true for some $n \geq 1$. Then

$$\sum_{i=1}^{n+1} i^2 = \sum_{i=1}^{n} i^2 + (n+1)^2 \qquad \text{peel off the last summand}$$

$$= \frac{n(n+1)(2n+1)}{6} + (n+1)^2 \qquad \text{by the inductive hypothesis}$$

$$= (n+1)\left(\frac{n(2n+1)}{6} + (n+1)\right) \qquad \text{factor out } n+1$$

$$= (n+1)\left(\frac{2n^2 + 7n + 6}{6}\right) \qquad \text{put over a common denominator}$$

$$= \frac{(n+1)(n+2)(2n+3)}{6} \qquad \text{factor}$$

Conclusion

The theorem is true when $n = 1$ and whenever it is true for some n, it is also true for $n+1$ $\big(P(n) \to P(n+1)$ is true$\big)$. The Mathematical Induction theorem implies that this theorem is true for all $n \geq 1$.

There is one more induction topic to discuss today. Some assertions are very difficult to prove using mathematical induction, but are fairly easy if we use a technique called *complete induction* (also called *strong induction*). Here is an example.

Theorem *The Fundamental Theorem of Arithmetic*

Every natural number $n \geq 2$ has a unique factorization as a product of primes in ascending order.

Consider proving the existence part of this theorem using mathematical induction. The inductive hypothesis would be: "assume that for some $n \geq 2$, n factors into a product of primes." How does knowing the factorization for n help in factoring $n + 1$? Knowing $8 = 2 \cdot 2 \cdot 2$ tells us nothing about the factorization of 9.

It's a mystery to me!

The Complete Induction Theorem provides a way out of this dilemma.

Theorem *Complete Induction*

If $\{P(n)\}$ is a set of statements such that

1. $P(1)$ is true   and

2. $[P(1) \wedge P(2) \wedge \cdots \wedge P(i)] \to P(i + 1)$ for $i \geq 1$

then $P(n)$ is true for all positive integers n. This can be stated more succinctly as

$$[P(1) \wedge (\forall i, [P(1) \wedge P(2) \wedge \cdots \wedge P(i)] \to P(i + 1))] \to [\forall n, P(n)]$$

In this form of induction, we assume more: namely that $P(n)$ is true for $1 \leq n \leq i$ and use some or all of those assertions to prove that $P(i + 1)$ is true.

We can use this theorem to prove the existence part of the Fundamental Theorem of Arithmetic. For the inductive step, we assume that for values of n with $2 \leq n \leq i$, n factors into a product of primes. Now consider $i + 1$. If $i + 1$ is a prime, then we have the desired factorization (as a product of one prime). Otherwise, $i + 1$ is composite, so we can find a and b with $1 < a < i + 1$ and $1 < b < i + 1$ such that $i + 1 = a \cdot b$. The inductive hypothesis implies that both a and b have prime factorizations, so $i + 1$ factors as the product of their factorizations. completing the inductive step.

85

Lily, I think it will be a good exercise for you to fully write up the complete induction proof I outlined.

You might think that complete induction allows you to prove more theorems than mathematical induction can be used to prove. That is not true. Anything that you can prove using one of the induction theorems can also be proved using the other. However, some theorems (such as the Fundamental Theorem of Arithmetic) might be easy to prove using complete induction and fiendishly difficult to do using mathematical induction.

Here is a final induction proof. It is about candy bars that are rectangular in shape but composed of a collection of joined squares. Two possible candy bar sizes are 4×1 and 3×5.

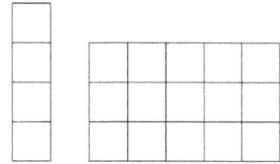

We can use complete induction to prove that it will always take $n - 1$ breaks to split the candy bar into n separate squares.

Base Step If the candy bar has only $n = 1$ square, then it takes $n - 1 = 0$ breaks to split it apart.

Inductive Step

Assume that any rectangular candy bar composed using k squares, with $1 \leq k \leq n$, can be split into k separate squares using exactly $k - 1$ breaks.

Now consider a candy bar composed of $n + 1$ squares. (Note that we can always build an $(n+1) \times 1$ candy bar.)

Using any of the break lines in the candy bar, break the bar into two rectangles having n_1 and n_2 squares, and $n + 1 = n_1 + n_2$. This requires one break. Since $1 \leq n_1 \leq n$ and $1 \leq n_2 \leq n$, the inductive hypothesis implies that the first square can be split using $n_1 - 1$ breaks and the second using $n_2 - 1$ breaks.

The total number of breaks employed was

$$1 + (n_1 - 1) + (n_2 - 1) = (n+1) - 1,$$

as desired.

Conclusion

The claim is true for $n = 1$ and whenever it is true for all k with $1 \leq k \leq n$, then it is also true for candy bars of size $n + 1$. The Complete Induction theorem implies that it is true for all candy bars of size n for $n \geq 1$.

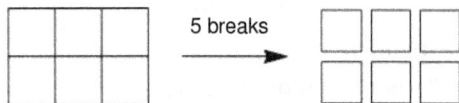

5 breaks

And that is the end of today's lesson. Spend some time reviewing this and trying some of the exercises.

Yes ma'am! This seems like it could be fun, once I master the two techniques.

Motivation for Induction

When a theorem makes a claim in the form "for all natural numbers n, *** is true," a proof using mathematical induction or complete induction should be considered. The proof outline looks like the following:

Base Step Hypothesis 1 of one of the induction theorems is verified.

Inductive Step Hypothesis 2 of the induction theorem is verified. Make sure you state the inductive hypothesis and document where you use it.

Conclusion This is where you show that you have satisfied the hypotheses of the induction theorem and can therefore assert that the claim is valid for all natural numbers greater than or equal to the initial value.

Theorems

Mathematical Induction If $\{P(n)\}$ is a set of statements such that

1. $P(1)$ is true   and
2. $P(i) \rightarrow P(i + 1)$ for $i \geq 1$

then $P(n)$ is true for all positive integers n. This can be stated more succinctly:

$$[P(1) \wedge (\forall i, P(i) \rightarrow P(i + 1))] \rightarrow [\forall n \geq 1, P(n)]$$

Complete Induction If $\{P(n)\}$ is a set of statements such that

1. $P(1)$ is true   and
2. $[P(1) \wedge P(2) \wedge \cdots \wedge P(i)] \rightarrow P(i + 1)$ for $i \geq 1$

then $P(n)$ is true for all positive integers n. This can be stated more succinctly:

$$[P(1) \wedge (\forall i, [P(1) \wedge P(2) \wedge \cdots \wedge P(i)] \rightarrow P(i + 1))] \rightarrow [\forall n, P(n)]$$

Partial Sum of a Geometric Progression Let $r \in \mathbb{R}$ with $r \neq 0$ and $r \neq 1$. Then

$$\sum_{i=0}^{n} r^i = \frac{1 - r^{n+1}}{1 - r} = \frac{r^{n+1} - 1}{r - 1}$$

The Fundamental Theorem of Arithmetic Every natural number $n \geq 2$ has a unique factorization as a product of primes in ascending order.

Exercises

Solutions can be found at `http://www.mathcs.bethel.edu/~gossett/DMGN/`.

1. Use mathematical induction to prove that for all natural numbers $n \geq 0$

$$\sum_{k=0}^{n} (2k + 1) = (n + 1)^2$$

2. Use mathematical induction to prove $2^n > n$ for all natural numbers $n \geq 1$.

3. Use mathematical induction to prove

$$\sum_{k=1}^{n} \frac{1}{2^k} = \frac{2^n - 1}{2^n} \quad \text{for } n \geq 1$$

4. Use mathematical induction to prove that $n^2 - 1$ is divisible by 8 for all positive odd integers n.

5. Use mathematical induction to prove that for all natural numbers $n \geq 1$

$$\sum_{k=1}^{n} k^3 = \frac{n^2(n + 1)^2}{4}$$

6. Use mathematical induction to prove that for all natural numbers $n \geq 1$

$$\sum_{k=1}^{n} \frac{k}{2^k} = \frac{2^{n+1} - n - 2}{2^n}$$

7. Use complete induction to prove (in full detail) the existence claim in the Fundamental Theorem of Arithmetic: every integer $n \geq 2$ has a factorization as a product of primes.

Chapter 7

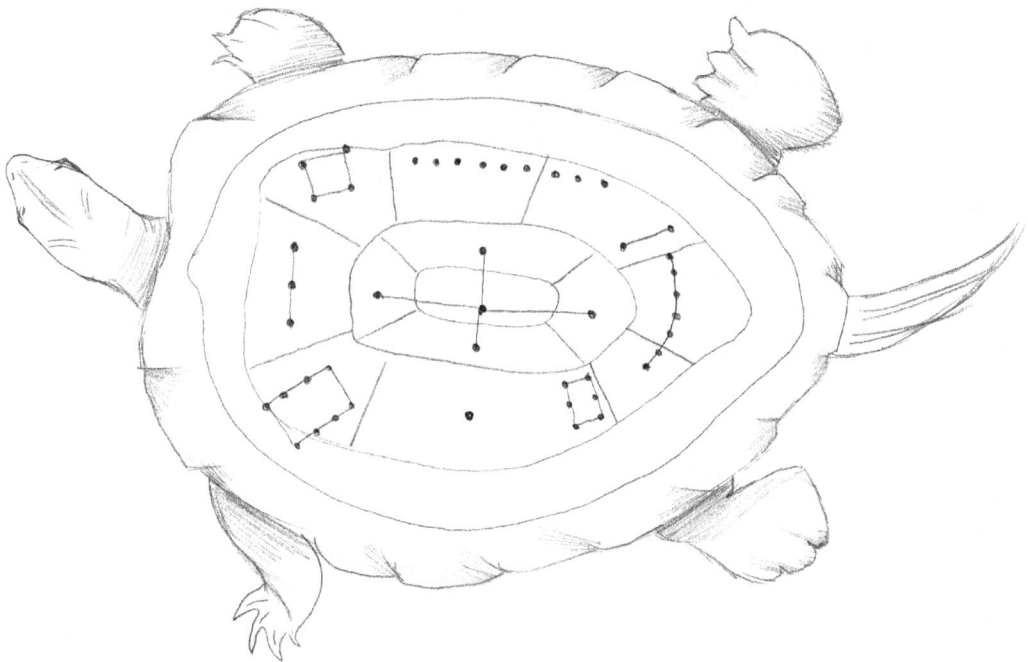

Isolde! I have a really fun puzzle for you! Logan learned about it in Chinese weekend school when he was younger. I brought him to tell you about it because he wouldn't tell me the solution.

Hi again Isolde. Lily insisted that I be the one to present this puzzle so she could learn the solution.

I am sure it is driving her crazy. So ... what is the puzzle?

The puzzle comes from an ancient Chinese legend. Long ago, central China was often devastated by huge floods. Finally Emperor Yu (大禹) devised an effective flood control system that diverted the waters into irrigation canals. The project took 13 years to complete (and many workers).

The legend concerns the river Lo. The people did not know how to appease the angry river god until a curiously marked turtle crawled up onto the bank. The turtle's shell contained the following diagram:

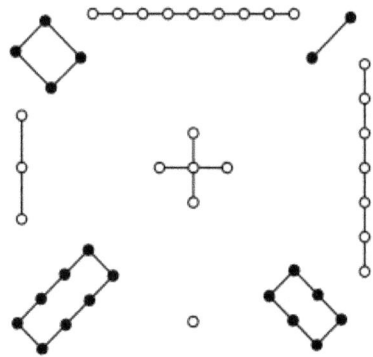

The diagram provided the clue to the number of sacrifices the people needed to offer to the river god. The diagram became known as the Lo Shu square: 洛書 luò shū (Lo Shu means the book of the river Lo).

The puzzle is to figure out what the diagram represents and why it might have been considered special.

Well, you gave me a clue when you said the diagram has something to do with the number of sacrifices. The simplest way to get numbers out of the diagram would be to count the circles. If I pay attention to the positions of the subfigures, I get the following arrangement:

$$
\begin{array}{ccc}
4 & 9 & 2 \\
3 & 5 & 7 \\
8 & 1 & 6
\end{array}
$$

This seems familiar.

Yes, you have made a very good start. What else do you observe?

I see that each of the integers from 1 to 9 is used exactly once. Oh! I know! It is a magic square.

What is a magic square?

A magic square is a matrix of n rows and n columns that consists of every number from 1 to n exactly once. In addition, the sum of every row, of every column, and the sum of the two diagonals are all the same number.

I see! The sums for the Lo Shu square are all 15. Are there other magic squares?

The Lo Shu magic square may be from around 2200 BC. In all that time people must have found others.

Actually, I recall something from my Math History course. Let me do a quick internet search.

Here it is. There is a record in India from around 1000 AD of a 4 by 4 magic square. It is

$$\begin{array}{cccc} 7 & 12 & 1 & 14 \\ 2 & 13 & 8 & 11 \\ 16 & 3 & 10 & 5 \\ 9 & 6 & 15 & 4 \end{array}$$

The common sum is 34.

So I was right! This was a very fun puzzle.

I know it is about time to start today's session, but I have a question before we start.
Logan is taking me to a movie tomorrow night. I was hoping you could come with us Isolde.

I don't want to intrude on your sibling time.

It is not an imposition. I would also enjoy having you come with us. In fact, I was hoping you could also join us for dinner before the movie. It would be my treat.

In that case, I would love to join the two of you tomorrow night.

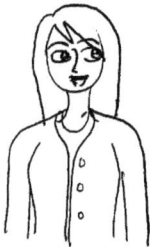

Great! Could you meet us here at 5 pm?

That works for me. Thanks again for the invitation.

Are you ready to start our tutoring session Lily?

Yes.

Counting—Part 1

Today we will start learning how to count.

I still don't see how this can be as difficult as you claim.

There is a lot more to counting than what you currently know, Lily. We won't even have time to learn all of it.

where did that pointer come from?

Here is a problem to think about. Suppose The Old Woman Who Lived In A Shoe has 20 children. She needs 4 of them to do laundry. How many different groups of size 4 can she form from her children?

So I just need to list all the teams of size 4. Sounds like a piece of cake if you ask me!

Spend some time right now to think about a complete solution.

94

Were you able to solve the problem I gave you to work on?

Yes, I'm pretty sure I solved it.

Can you show me your solution?

Sure!

I first named the children as a, b, c, ..., r, s, t. I then started to systematically list all the subsets of size 4.

abcd, abce, abcf, ..., abct
bacd, bace,   ...

But then I noticed that abcd and bacd were really the same set of 4 children.

So I made sure the two rows have no duplicates by omitting "a" from the second row. I used the same idea for the third row and the following rows.

abcd, abce, abcf, ..., abct
bcde, bcdf, ..., bcdt
cdef, cdeg, ..., cdet
$$\vdots$$
pqrs, pqrt
qrst

Then I counted the number of subsets in each row and added them:

$$17 + 16 + 15 + \cdots + 3 + 2 + 1 = 153$$

It's really good that you noticed the duplicate subsets.

But I know that your answer is incorrect. You missed lots of subsets. "a" only appears in your first row, but every time it appears, so do "b" and "c". So you missed all subsets with "a" but not "b."

How could I have missed those? Let's write them on the board properly.

You need to replace your first row with:

abcd, abce, abcf, . . . , abct
abde, abdf, abdg, . . . , abdt
abef, abeg, . . . , abet
⋮
abst
acde, acdf, . . . , acdt
⋮

And then do something similar for your other rows. That is going to be more complicated.

Ugh! Is there an easier way to count this?

Yes! But first we need to discuss some other ideas.

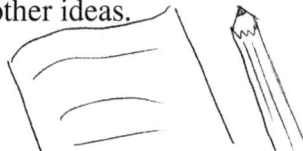

Let's start with a new problem. Suppose tonight you can either read a book, watch a movie at home, or work on homework.

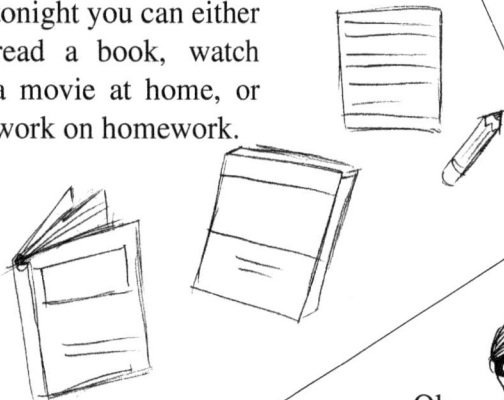

Suppose that you also want to eat a snack. The snack choices are: popcorn, corn chips, toast, and rice crackers. If the choice of activity has no influence on the choice of snack, we say that the two choices are *independent*.

Ok.

We can make a table for the choices. Just place an X in the box that represents your choice.

If you want to read a book and eat chips, you would mark it like this.

Yes. That makes sense.

So here is our first counting principle:

The Independent Tasks Principle

If a project can be decomposed into two independent tasks with n_1 ways to accomplish the first task and n_2 ways to accomplish the second task, then the project can be completed in $n_1 \cdot n_2$ ways.

To use this, you just need to be careful that the two tasks are independent. For example, if the tasks are choosing a skirt and choosing a shirt, you will probably not consider them to be independent. A red skirt with a green shirt is not a good idea unless you are going to a Christmas party.

Wouldn't this work with multiple independent tasks? Suppose I want to decorate cookies. With 2 kinds of cookies, 3 kinds of frosting, and 4 kinds of candy toppings, there are $2 \cdot 3 \cdot 4 = 24$ ways to make a cookie.

Very good Lily! You are getting the hang of this. I think you are ready for the next counting principle.

The Mutually Exclusive Tasks Principle

If a project can be accomplished by completing exactly one of two mutually exclusive tasks, with n_1 ways to accomplish the first task and n_2 ways to accomplish the second task, then the project can be completed in $n_1 + n_2$ ways.

You need to be careful that the two tasks are mutually exclusive. For example, if the tasks are choosing one of 3 romance novels, or choosing one of 4 fantasy novels to read, there may be novels that fit both categories. Using this principle would cause you to incorrectly count such a novel twice.

Suppose on Saturday you can either visit one of 3 friends, or you can do one of 4 household chores. If you choose to visit Mary, you won't be able to do any household chores. If you decide to vacuum, you won't be able to visit a friend. We call such sets of choices *mutually exclusive*. Choosing an option from one set automatically excludes choosing an option from the other set.

I think I get it. I could put the names of the 3 friends in a hat. I could then write the 4 chores on different slips of paper and add those to the hat. I make my choice by choosing one slip of paper. So I can choose in $3 + 4 = 7$ ways.

Correct!

We can use these two counting principles in many situations, but they don't always apply. Can you think of two tasks that are neither independent nor mutually exclusive?

I think so. When I read I like to listen to music. I could let task 1 be "choose a book" and task 2 be "choose a CD."

These aren't mutually exclusive because I pick something from each category.

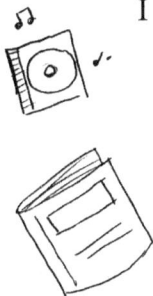

They aren't independent because when I am reading a textbook, music with words is distracting, so some CDs are not viable options with that choice of book.

That is a great example. In a situation like that we usually just list all the acceptable pairs of choices and count the list.

Permutations

There is a special case of the independent tasks principle that is so common it has been given a name and its own counting formulas.

Suppose you wish to phone 4 friends. If the order of the calls is important, in how many ways can you complete the calls?

#	1st call	2nd call	3rd call	4th call
P	Mary			

Well, I have 4 choices for the first call. Once that is done, I still have 3 friends left to call.

Who I call next only depends on who is left, so which of the three remaining friends I chose will be independent of the completed first choice.

#	1st call	2nd call	3rd call	4th call
P	Mary	Jane		

I then have 2 choices for the third call and only one choice for the final call. The independent tasks principle asserts there are $4 \cdot 3 \cdot 2 \cdot 1 = 24$ ways to arrange the calls.

#	1st call	2nd call	3rd call	4th call
P	Mary	Jane	Emma	Kate

Good job!

There is a notation, called *factorial notation*, that is very helpful for counting. Your last calculation can be abbreviated as $4 \cdot 3 \cdot 2 \cdot 1 = 4!$.

The "exclamation point" is read as "factorial." In general, we define $n!$ by

$$n! = n \cdot (n-1) \cdot (n-2) \cdots 3 \cdot 2 \cdot 1$$

where the $\cdots$ collapses if n is 3 or less.

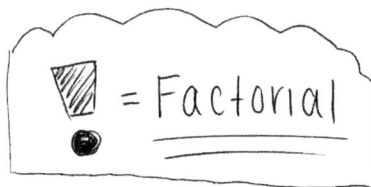

= Factorial

So:

$$4! = 4 \cdot 3 \cdot 2 \cdot 1 = 24$$
$$3! = 3 \cdot 2 \cdot 1 = 6$$
$$2! = 2 \cdot 1 = 2, \text{ and}$$
$$1! = 1$$

The mathematicians around the world got together and decided to define $0! = 1$. One of the many reasons why this is the proper definition is to make the counting formula $C(n, n)$ equal 1, as we expect. (See the definition on page 105.)

$$0! = 1$$

Here is a new problem. If 8 people enter a piano competition and there are 3 prizes (1st, 2nd, and 3rd), in how many ways can the judges award the three prizes? (Assume there are no ties.)

This is similar to the last problem. There are 8 choices for 1st prize, 7 choices left for 2nd prize, and 6 choices left for 3rd prize. So there are

$$8 \cdot 7 \cdot 6 = 336$$

ways to award the prizes.

Correct again. Here is a way to summarize how to count when order matters but repetition of choices is not allowed.

Suppose we start with n options and want to choose k of them (with $1 \leq k \leq n$) and order matters. Denote the number of ways to do this as $P(n, k)$. Then

$$P(n, k) = n(n - 1)(n - 2) \cdots (n - (k - 1))$$
$$= n(n - 1)(n - 2) \cdots (n - k + 1)$$

Why do we quit at $n - (k - 1)$? Oh, wait. I see. We start counting at $n = n - 0$ and want k items, so the kth number to subtract from n is $k - 1$. In the piano contest $n = 8$ and $k = 3$, so $k - 1$ is 2 and

$$P(8, 3) = 8 \cdot 7 \cdot 6$$
$$= (n - 0)(n - 1)(n - 2)$$
$$= n(n - 1)(n - 2)$$

Notice that

$$P(n, k) = n(n - 1)(n - 2) \cdots (n - (k - 2))(n - (k - 1))$$

$$= \frac{n(n-1)(n-2) \cdots (n-(k-2))(n-(k-1))(n-k)(n-(k+1)) \cdots (3)(2)(1)}{(n-k)(n-(k+1)) \cdots (3)(2)(1)}$$

$$= \frac{n!}{(n - k)!}$$

since the product in the denominator cancels with everything after $(n - (k - 1))$ in the numerator. This can be stated as our next counting formula.

Permutations

Let $0 \leq k \leq n$. The number of ways to arrange k objects from a set of n objects, in order, but without repetition is

$$P(n, 0) = 1$$

$$P(n, k) = n \cdot (n - 1) \cdot (n - 2) \cdots (n - k + 1) = \frac{n!}{(n - k)!}$$

for $k \geq 1$

The order in which the objects are arranged is important. Also, once an object is used, it cannot be used again.

Here is another variation, called *permutations with repetition*. Suppose you want to create a 5-digit postal code. How many distinct postal codes will there be? Notice that order matters: 55112 and 11255 are distinct.

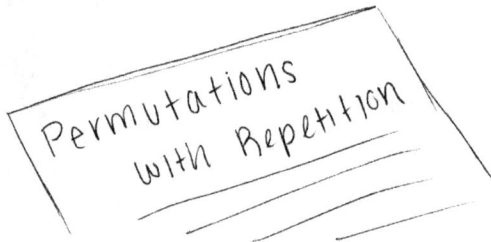

Permutations with Repetition

The independent tasks principle works again. There are 10 digits (0, 1, 2, 3, 4, 5, 6, 7, 8, 9), so there are 10 choices for the first digit and 10 for each of the remaining digits. So there are

$$10 \cdot 10 \cdot 10 \cdot 10 \cdot 10$$
$$= 10^5 = 100,000$$

possible postal codes.

That's right. So if we have n options and want to choose k of them, in order, but with repetition allowed, there are n^k ways to do this.

What if order doesn't matter? I didn't worry about order for the first problem we did (with the Old Woman Who Lived In A Shoe). I just created groups of 4 to do the laundry.

Good observation! It leads into our next topic: *combinations*.

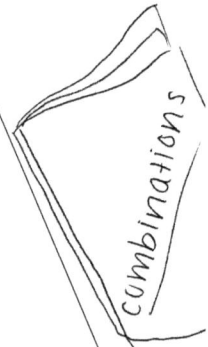

combinations

Combinations

The laundry problem is an example of a combination without repetition. The key feature is that order *doesn't* matter.

A combination is like a committee. In many committees everyone is equal; there is no implied order or ranking of the members.

Combinations are subsets. By definition, sets (and hence subsets) are always unordered. (See page 12.)

The Gerbil Welfare Committee

How do we count combinations? Are they related to combination locks?

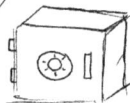

Actually, the combination for a combination lock is really a permutation (since the correct order is essential). I bet combination locks were *not* named by a mathematician.

Wow, this is starting to get confusing again ...

Let's think about a simpler combinations problem.

Suppose 4 friends want to have a movie night. They each bring a favorite movie, but only have time to watch 3 of the movies. If they don't care about the order in which the movies are viewed, in how many ways can they choose a set of 3 to watch?

This is simple enough that we can list all potential sets in a systematic manner. Let's denote the 4 movies by A, B, C, and D and not worry about order just yet.

ABC	ABD	ACD	BCD
ACB	ADB	ADC	BDC
BAC	BAD	CAD	CBD
BCA	BDA	CDA	CDB
CAB	DAB	DAC	DBC
CBA	DBA	DCA	DCB

There are 24 entries.

Now we pay attention to order. Observe that all the entries in the first column represent the same set of three movies. So we have counted each choice 6 times. Notice also that the entries in column 1 are all $3! = 6$ permutations of the movies A, B, and C. Similar observations apply to the other columns. So we really have just $24/6 = 4$ choices.

ABC	ABD	ACD	BCD

I get it! We count the permutations using $P(n, k)$ then divide by $k!$ (where k is the number of items in the subset).

But I have an easier way to count for this problem: just count the number of ways to leave out 1 movie. There are 4 movies, so 4 ways to skip one movie.

That's great Lily! Hold that thought. You have discovered a special case of a nice theorem.

The ideas in the movie problem apply in all combinations problems. So the counting formula for combinations is:

Combinations

Let $0 \le k \le n$. The number of ways to choose a subset of k objects from a set of n objects without repetition is

$$C(n, k) = \frac{n!}{k! \cdot (n - k)!}$$

If $0 \le n < k$, then $C(n, k) = 0$. $C(n, k)$ is also denoted as $\binom{n}{k}$.

So the answer to the problem with 20 kids where 4 must do the laundry is

$$C(20, 4) = \frac{20!}{4! \cdot 16!} = \frac{20 \cdot 19 \cdot 18 \cdot 17 \cdot \cancel{16} \cdot \cancel{15} \cdots \cancel{2} \cdot \cancel{1}}{(4 \cdot 3 \cdot 2 \cdot 1)(\cancel{16} \cdot \cancel{15} \cdots \cancel{2} \cdot \cancel{1})} = \frac{20 \cdot 19 \cdot 18 \cdot 17}{4 \cdot 3 \cdot 2 \cdot 1} = 4845.$$

I don't think I would have ever listed all of them correctly.

Well, that's about all we have time for today. You did really well for our first counting lesson, Lily.

Thanks, Isolde! This was really fun!

I'm glad it wasn't too much of a challenge for you. Before you go I have a homework assignment for next week.

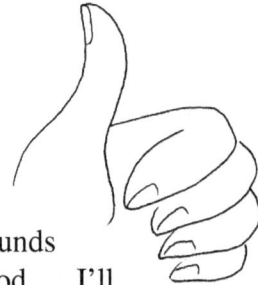

Sounds good. I'll give it my best shot!

Lily's Assignment to Complete Before Counting – Part 2:

1. Find or create an interesting combinations problem.

2. When discussing the movie night problem, you correctly claimed that $C(4,3) = C(4,1)$. Prove the general version of your theorem:

$$C(n,k) = C(n,n-k)$$

Here is a summary of the counting formulas presented so far:

The Independent Tasks Principle

If a project can be decomposed into two independent tasks with n_1 ways to accomplish the first task and n_2 ways to accomplish the second task, then the project can be completed in $n_1 \cdot n_2$ ways.

The Mutually Exclusive Tasks Principle

If a project can be accomplished by completing exactly one of two mutually exclusive tasks, with n_1 ways to accomplish the first task and n_2 ways to accomplish the second task, then the project can be completed in $n_1 + n_2$ ways.

	Permutation With Order	**Combination** Without Order
Without Repetition	$P(n,k) = \frac{n!}{(n-k)!}$	$C(n,k) = \binom{n}{k} = \frac{n!}{k! \cdot (n-k)!}$
With Repetition	n^k	

Table 1: Arranging k elements from a set containing n distinct elements.

Exercises

Solutions can be found at `http://www.mathcs.bethel.edu/~gossett/DMGN/`
but solve them yourself before checking the answers. Before solving, identify the counting formula or formulas that you will use, and why they are appropriate for the exercise.

1. How many license plates can be created using the form DLLDDD where D is a digit and L is an uppercase letter from the English alphabet?

2. A store uses two kinds of codes to represent their inventory. The code for **N**on-taxable items is NDDD and the code for **T**axable items is TDDDD, where N and T are those specific letters and D represents any digit. How many different items can they encode with their system?

3. A 3rd grade teacher has 22 students and 25 desks. In how many ways can she create a seating chart for the students?

4. A lazy teacher has decided to create a 15-question true-false exam. He plans to create the answer key first, and then design questions to fit the answer key. In how many ways can he create the answer key?

5. I have 8 pots of tulips. Three of the pots have blue flowers and 5 have red flowers. In how many visually distinguishable ways can I arrange the pots in a line? Assume that all the tulips are the same size and the pots are identical. The colors red and blue are the only distinguishable features; all the red tulips are identical and all the blue tulips look the same as well.

6. A teacher has 12 boys and 14 girls in her class. She wants to form a committee of 14 students. At least half of the 12 boys and at least half of the 14 girls must be on the committee. In how many ways can she form the committee?

7. A combination lock has a dial with 36 choices. If the combination sequence uses 3 numbers, how many combination sequences are possible?

8. Tonight I will either read one of 2 novels or watch one of 3 DVDs. In addition, I will choose 2 of 6 possible snacks to eat while reading or viewing a DVD. In how many distinct ways can I make these choices?

9. A teacher has 23 students in his class. The class will stage a short play next week. There are 3 main parts (the narrator, a tooth, and a toothbrush). There will also be 5 students chosen as dental cavities (after the three main leads have been chosen). In how many ways can the parts be assigned if there is no distinction made among the cavities?

10. Ignacio plans to eat at a restaurant tonight. He can either go by himself or invite one or more of his five friends. Assuming that different mixes of friends create distinct dining experiences, in how many ways can Ignacio plan his restaurant event?

Help Lily find the correct solution to the counting problem.

Frostbite Falls Elementary School has four fifth grade classes. Each of the classes has 25 students enrolled. The students in each class must choose three students to compete in a sack race. The twelve competitors will receive ribbons (first place through twelfth place). In how many ways can the ribbons be awarded?

Solution 1

1. Choose the 12 students. Since this will be a group of competitors, we need a combination.
$$\binom{100}{12} = 1050421051106700 \text{ ways}$$

2. Determine in how many ways the 12 students can receive ribbons.
$$12! = 479001600 \text{ ways}$$

3. These two steps are independent, so multiply to get the final answer.
$$\binom{100}{12} \cdot 12! = 503153364153791070720000 \text{ ways}$$

Solution 2

1. Choose 3 students from each class. Since this will be a group of competitors, we need a combination.
$$\binom{25}{3} = 2300 \text{ ways}$$

2. The choices in a class are independent of the choices in other classes, so the four sets of choices are independent. The number of ways to pick the 12 competitors is:
$$\binom{25}{3}^4 = 27984100000000$$

3. Determine how many ways the 12 students can receive ribbons.
$$12! = 479001600 \text{ ways}$$

4. The previous two steps are independent, so multiply to get the final answer.
$$\binom{25}{3}^4 \cdot 12! = 13404428674560000000000 \text{ ways}$$

Chapter 8

Hello Isolde. How is the tutoring job progressing? I hear good things from Lily's parents.

Good morning, professor Douglas. I think it is going well. I am really enjoying the experience. Lily is charming, and she is also very bright.

I have a question, if you have time. I am planning to give Lily some study hints for learning mathematics. Do you have any suggestions?

I do have a few ideas I share with my students. I need to get to a class right now. Why don't you come to my office later today and I can share them with you.

Later that day.

Hello again Isolde. I have time to talk now.

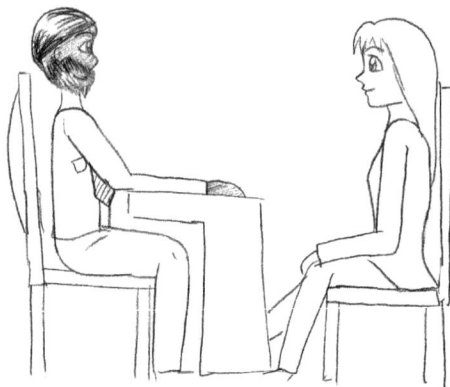

I haven't prepared anything systematic. I imagine you can find such presentations in your education classes.

I think one of the biggest issues in learning anything is how much time is spent on the task. Many students think they can learn a subject by spending only 30 minutes a day outside of class. Unfortunately, there is no way that real learning can happen with such a tiny investment of time.

For a university-level course, you may need to spend 2-4 hours outside of class for each hour in class.

Here are a few other suggestions.

(1) Make sure you are enrolled in the proper course. If you don't have the proper background, you will find the course overwhelming. In particular, many students experience a challenging transition from high school courses to university courses. Be ready to meet the challenge.

(2) Start homework assignments early. If class is on Monday and the homework is due on Wednesday, start working on it Monday night. That way, if you get stuck you have time to get help. If you start at 2 a.m. Wednesday morning and class is at 8 a.m., you have few options for help if you don't understand something.

(3) Read the textbook slowly and carefully *before* you start the exercises. Read with paper and pencil. Actively work through the examples in the book. Don't just passively assume that it makes sense.

(4) If you are stuck on an exercise for more than 15 minutes, (a) make sure you understood the reading—review the relevant parts, (b) move on to the next exercise and plan to get help on the stalled exercise.

If you are stalled on several exercises, move on to a different subject. Go to office hours or the math tutor lab or meet with a classmate. Determine why you are stuck and master the missing ideas.

Should students work on their own, or form study groups?

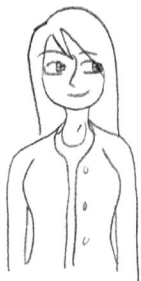

Working with a classmate (or perhaps a few others) can be better than working alone, if it is done properly. Here are some guidelines:

- The students should be sharing (and discussing) ideas. Acting like a human copy machine is cheating, not studying.

- The group should make sure that everyone understands the ideas before moving on.

- Groups work best if there are no large disparities in current abilities.

- In my classes, I tell the students that they are encouraged to collaborate on homework, but they should write the exercise solutions on their own, after the group session is over. That way, what is written reflects what is left in their own brain and is not a copy of someone else's ideas. If the idea is still there after they split up, then I assume that the idea is now something they have mastered.

What should the student do if all these suggestions are not sufficient?

I would suggest going to talk with the instructor as a first step. Students often don't believe this, but their instructors really do want them to succeed. If the instructor thinks it would be helpful, they might look for a private tutor.

Thanks for your time, professor Douglas. This has been very helpful. Do you have any other final words of advice for students?

Get more sleep. University students are notorious for their lack of sleep. Some essential processing of information occurs in the brain while you sleep. Depriving yourself of sufficient sleep means those essential steps are missed, causing those extra hours of studying to be wasted. You will have greater knowledge retention if you sleep more and do less fruitless cramming.

Counting—Part 2

$$\binom{n+m}{k} = \sum_{i=0}^{k} \binom{n}{i}\binom{m}{k-i}$$

$$\binom{2+3}{2} = \sum_{i=0}^{2} \binom{2}{i}\binom{3}{2-i}$$

$$= \binom{2}{0}\binom{3}{2} + \binom{2}{1}\binom{3}{1} + \binom{2}{2}\binom{3}{0}$$

$$n = 2:\ \{A, B\} \qquad m = 3:\ \{a, b, c\}$$

$\binom{2}{0}\binom{3}{2}$ $\qquad\qquad$ $\{a,b\}$ $\quad$ $\{a,c\}$ $\quad$ $\{b,c\}$

$\binom{2}{1}\binom{3}{1}$ $\quad$ $\{A,a\}$ $\ $ $\{A,b\}$ $\ $ $\{A,c\}$ $\ $ $\{B,a\}$ $\ $ $\{B,b\}$ $\ $ $\{B,c\}$

$\binom{2}{2}\binom{3}{0}$ $\qquad\qquad\qquad$ $\{A, B\}$

How did the homework from last week go for you?

It went very well. I'm sure there are no mistakes this time.

You said that I could find an example of a combinations problem, so I asked Logan to help me come up with one.

Can you show me the combinations problem on which you both worked?

Logan is in a software engineering class. In *The mythical man-month*, Frederick Brooks discusses communication on a team project. If there are two people, then there is only 1 communication link.

If the team has three people, there will be 3 links (one for each pair of people).

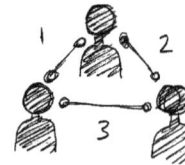

With 4 people there are 6 links. If the team consists of n people, then there will be

$$C(n, 2) = \frac{n!}{2! \cdot (n-2)!} = \frac{n(n-1)}{2}$$
$$= \frac{1}{2}n^2 - \frac{1}{2}n \text{ links.}$$

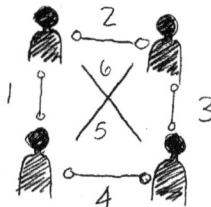

His point is that as new people are added to the project, the number of communication links grows faster than the team size grows. So an organizational structure is needed to reduce the number of communication links. Like having mid-level managers and sub-teams. The people on a sub-team don't usually need to talk directly with everyone on some other sub-team.

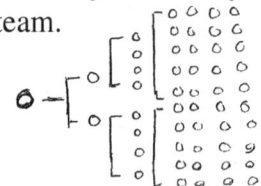

That's great. Do you also have a proof for the counting theorem?

Of course!

I worked on this for a while by myself. Then I realized it's really a proof by "crossing your eyes." It looks like this:

cross your eyes

$$C(n,k) = \frac{n!}{k! \cdot (n-k)!} \overset{\downarrow}{=} \frac{n!}{(n-k)! \cdot k!}$$

$$= \frac{n!}{(n-k)! \cdot (n-(n-k))!}$$

$$= C(n, n-k)$$

Very nice! Let me introduce a convenient alternative notation. $C(n,k)$ is often written as $\binom{n}{k}$ and it is pronounced as "n choose k." This is called binomial notation.

Ok. But why do we need another notation?

We don't *need* another notation, but people have found this other notation to be useful—especially in long calculations.

$$\binom{n}{k} = C(n,k)$$

Here is our first new topic for today. Suppose you have 4 identical black marbles, 6 identical white ones, and 5 identical gray marbles. In how many visually distinguishable ways can you place all 15 marbles in a line?

Visually distinguishable?

OR

I will try to explain what visually distinguishable means.

Suppose I have 2 black and 1 white marble. Denote the black marbles as B_1 and B_2 and the white marble as W. Then if the black marbles are identical, the arrangements $B_1 W B_2$ and $B_2 W B_1$ are visually *indistinguishable*. However, the arrangements $B_1 W B_2$ and $B_1 B_2 W$ would be visually *distinguishable*.

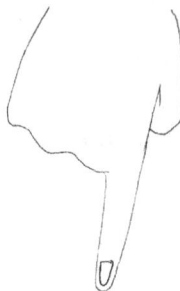

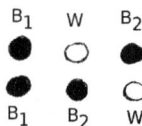

B_1 $\quad$ W $\quad$ B_2

B_1 $\quad$ W $\quad$ B_2

B_1 $\quad$ B_2 $\quad$ W

Ok. I now know what visually distinguishable means, but I am not sure how to solve the original question. Does it involve combinations?

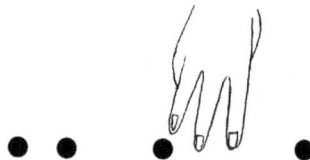

Yes. We can approach the original problem as a sequence of three independent choices. (1) Decide which of the 15 positions will contain the 4 black marbles.

(2) Decide which of the 11 remaining positions should contain the 6 white marbles.

(3) Decide which of the 5 remaining positions should contain the 5 gray marbles (all of them).

I get it. Since they are independent tasks, we multiply the number of ways to accomplish the three tasks. When I write it out, there is a lot of cancellation.

$$\binom{15}{4} \cdot \binom{11}{6} \cdot \binom{5}{5} = \frac{15!}{4! \, 11!} \cdot \frac{11!}{6! \, 5!} \cdot \frac{5!}{5! \, 0!} = \frac{15!}{4! \, 6! \, 5! \, 0!}$$

118

Wow! You picked up the new notation very quickly. Remember that $0! = 1$, so the answer can be written more simply as

$$\frac{15!}{4! \, 6! \, 5!}$$

My calculator indicates there are 630,630 ways to line up the marbles.

I bet there is nothing special about the numbers of marbles. I think it also doesn't matter much that there were exactly three colors.

Correct. Here is our next counting theorem.

The Multinomial Counting Theorem

Suppose there exists a set of n items containing n_1 *identical* items of type 1, n_2 *identical* items of type 2, ..., and n_k *identical* items of type k, where $n = n_1 + n_2 + \cdots + n_k$. The number of visually distinguishable ways to arrange the n items in a row is

$$\text{multinomial}(n_1, n_2, \ldots, n_k) = \frac{n!}{n_1! \cdot n_2! \cdots n_k!}$$

The proof is essentially the same as what we did with $n_1 = 4, n_2 = 6, n_3 = 5$.

Do you understand this theorem? There are two more counting ideas to look at if you are ready.

The multinomial counting theorem seems pretty easy to use. Let's move on.

Ok. The next idea is counting by setting up a one-to-one correspondence. Have you ever gone to summer camp at a lake?

Yes. The water was very cold, so I only swam on very hot days.

One year I was a lifeguard at a summer camp.

To ensure that every swimmer was safe, we made them swim with a partner. When a lifeguard blew a whistle, each swimmer needed to hold up the hand of their partner.

1-1 correspondences designate partners between the objects in two sets. Every member of each set has exactly one partner.

That means the two sets are the same size. For example, if I look at a classroom that has 20 desks and discover that there are no empty desks and no students that are standing, then I know there are 20 students.

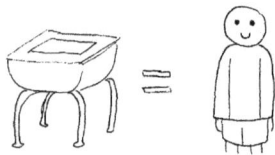

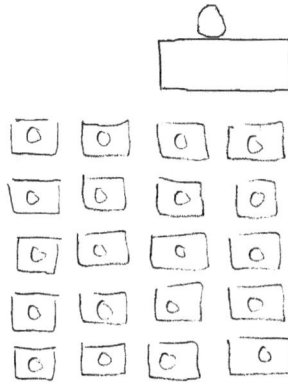

Alright, I think I'm following so far.

Here is where things get a bit weird. We can sometimes set up a 1-1 correspondence between two infinite sets.

Here's an example. Let $\mathbb{N}$ be the set of natural numbers and E be the set of even natural numbers. Every number in E is also in $\mathbb{N}$, but there are numbers (such as 7) in $\mathbb{N}$ that are not in E. However, we can set up a 1-1 correspondence between the two sets, so they must actually be the same size. (Both countably infinite.) Here is the correspondence. List each number in $\mathbb{N}$ in the top row and its partner in E directly below.

$$\begin{array}{c|ccccc} \mathbb{N} & 0 & 1 & 2 & 3 & 4 & \cdots \\ E & 0 & 2 & 4 & 6 & 8 & \cdots \end{array}$$

Every number in each set has exactly one partner, so the sets must be the same size, even though we intuitively feel that E is smaller.

This 1-1 correspondence isn't guaranteed. We can place the natural numbers and the rational numbers (the set of all fractions) in one-to-one correspondence, but we cannot do that with the natural numbers and the real numbers. So the set of real numbers is an infinite set that is larger than countably infinite. I don't intend to prove these claims right now, but the proofs are not difficult (see pages 132 and 133).

Is all of this still making sense Lily?

Yes, I'm following everything. But I have a question about last week's material.

combinations with repetition

Last week we discussed permutations with and without repetition. However, you only talked about combinations *without* repetition. Will we ever cover combinations *with* repetition?

Actually, that is what I plan to discuss next. I have a very sweet example to introduce the idea.

Suppose we go to the donut store. The store has 5 varieties on hand. In how many ways can we choose a dozen donuts? You may assume that all donuts of the same variety are identical.

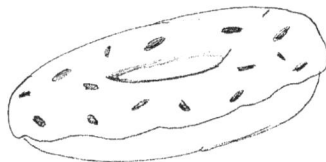

This sounds a bit like the multinomial counting theorem. Could I use it with n_i being the number of donuts of variety i?

That's a good guess Lily, but you haven't counted all the options using that approach. See if you can figure out your mistake.

Ok. I think I have it ... We need to count not just for $n_i = 3$ glazed donuts, but also for n_i being *any* value in $\{0, 1, 2, \ldots, 12\}$. And we need to do this for all 5 values of i.

That's right. However, we can make a clever observation that allows us to use the multinomial theorem in a correct manner.

Suppose that each donut costs 25 cents. You have 12 quarters in your pocket. There are five varieties of donuts, so if you start with variety 1, you need to decide how many of that variety to choose and then make a decision to move on to variety 2. In the same way, you choose some donuts of variety 2 (including choosing 0 donuts of that variety) and then you need to decide to move to variety 3. In all, you make 12 donut choices and 4 decisions to move on. We can represent a donut choice by Q (for the quarter that will be spent), and each change in variety by an N (for Next variety). If we line these 16 letters up, we can uniquely specify our mix of donuts.

For example, the line QQNNQQQQQNQNQQQQ corresponds to the selection $v_1 v_1 \ v_3 v_3 v_3 v_3 v_3 \ v_4 \ v_5 v_5 v_5 v_5$.

The selection $v_1 v_1 v_1 \ v_2 v_2 v_2 \ v_3 v_3 v_3 \ v_4 v_4 v_4$ corresponds to the letter pattern QQQNQQQNQQQNQQQN. So there is a 1-1 correspondence between patterns of Qs and Ns and unique donut selections.

To count the number of ways to select a dozen donuts we just need to count the number of visually distinguishable patterns of 12 Qs and 4 Ns. Consequently, there are

$$\text{multinomial}(12, 4) = \frac{16!}{12! \cdot 4!} = C(12 + (5 - 1), 12) = 1820$$

possible selections of donuts.

5 varieties   12 donuts
QQNNQQQQQNQNQQQQ
corresponds to
$v_1 v_1 \ v_3 v_3 v_3 v_3 v_3 \ v_4 \ v_5 v_5 v_5 v_5$.

The number of Ns is one less than the number of varieties. The number of Qs is equal to the number of donuts to choose.

So the Ns represent a change in variety and the Qs represent donuts (the things we wish to select). Thus Q is sort of like k in $C(n, k)$. I think I can guess the general pattern.

Combinations With Repetition

Let there be $n \geq 1$ categories of items, and let $k \geq 0$. If each category contains at least k copies of its item, then the number of distinct subsets of k elements (with repetition permitted), is

$$C(k + (n-1), k) = C(n + k - 1, k) = \frac{(n + k - 1)!}{k! \cdot (n-1)!}$$

Ok. That fills in the rest of the table from last time.
So now I know all the counting techniques.

I'm afraid not. There are many counting techniques we haven't discussed. For example, the pigeon-hole principle, inclusion-exclusion, container problems, counting using generating functions, and others.

Wow! Are we going to do all that?

We may not have time to explore those topics.

I do have one final problem for you this week.
The solution uses several of the counting techniques you have already learned.

	Permutation With Order	Combination Without Order
Without Repetition	$P(n,k) = \frac{n!}{(n-k)!}$	$C(n,k) = \frac{n!}{k! \cdot (n-k)!}$
With Repetition	n^k	$C(n+k-1,k) = \frac{(n+k-1)!}{k! \cdot (n-1)!}$

Arranging k elements from a set containing n distinct elements.

Independent Tasks Principle

$n_1 \cdot n_2 \cdot n_3 \cdots n_k$

Mutually Exclusive Tasks Principle

$n_1 + n_2 + n_3 + \cdots + n_k$

Ok. Sounds fun.

Here is the problem. A high school math team has 4 seniors and 5 juniors. The coach needs a traveling team of 5 students for a contest in another city. The coach has decided that the traveling team must contain at least 2 seniors. In how many ways can the traveling team be formed?

Seniors   Juniors

Some time passes as they both work on a solution . . .

I have a solution. The coach can first choose 2 seniors (ensuring that the team has at least 2 seniors). Then she can choose the remaining 3 competitors from the 7 remaining candidates. These choices are independent, so multiply:

$$\binom{4}{2}\binom{7}{3} = 210 \text{ possible traveling teams.}$$

I used the independent tasks formula and the combinations formula.

Here is my solution. Teams with exactly 2 seniors, exactly 3 seniors, and exactly 4 seniors are mutually exclusive choices. If I have exactly 2 seniors, then there are 5 juniors from which to choose the remaining competitors. So (also using combinations and independent tasks), I get

$$\binom{4}{2}\binom{5}{3} + \binom{4}{3}\binom{5}{2} + \binom{4}{4}\binom{5}{1} = 105 \text{ possible traveling teams.}$$

So my solution is wrong.

Not necessarily. We could both be wrong. Otherwise either you over-counted or I under-counted.

Hmm. Let me consider an example. Let the seniors be A, B, C, and D and let the juniors be V, W, X, Y, and Z.

If the traveling team is ABC VW then I could pick that set as A,B and then C, V, W or as B, C and then A V, W. So I definitely over-counted.

Yes, you did. But is my solution correct? One way to check is to count using a different strategy and see if I get the same solution.

To check your answer, try to use a different solution technique.

Here is an idea. There are $\binom{9}{5} = 126$ ways to pick a traveling team with no restrictions on membership. Let's subtract from this the number of ways to form a traveling team that doesn't contain at least 2 seniors.

There is only 1 team with no seniors: VWXYZ. There are 20 teams that contain exactly one senior:

	include A	include B	include C	include D
omit Z	AVWXY	BVWXY	CVWXY	DVWXY
omit Y	AVWXZ	BVWXZ	CVWXZ	DVWXZ
omit X	AVWYZ	BVWYZ	CVWYZ	DVWYZ
omit W	AVXYZ	BVXYZ	CVXYZ	DVXYZ
omit V	AWXYZ	BWXYZ	CWXYZ	DWXYZ

So there are $126 - 1 - 20 = 105$ possible traveling teams, which agrees with my previous answer. The two results can be summarized as:

$$\binom{4}{2}\binom{5}{3} + \binom{4}{3}\binom{5}{2} + \binom{4}{4}\binom{5}{1} = \binom{9}{5} - \binom{4}{0}\binom{5}{5} - \binom{4}{1}\binom{5}{4}$$

I think I can make a nicer expression:

$$\binom{9}{5} = \binom{4}{0}\binom{5}{5} + \binom{4}{1}\binom{5}{4} + \binom{4}{2}\binom{5}{3} + \binom{4}{3}\binom{5}{2} + \binom{4}{4}\binom{5}{1}$$

Actually, you have discovered an example of Vandermonde's Theorem. If we make the agreement that $\binom{n}{k} = 0$ whenever $k > n$, then we can write your equation as

$$\binom{9}{5} = \binom{4}{0}\binom{5}{5} + \binom{4}{1}\binom{5}{4} + \binom{4}{2}\binom{5}{3} + \binom{4}{3}\binom{5}{2} + \binom{4}{4}\binom{5}{1} + \binom{4}{5}\binom{5}{0} = \sum_{i=0}^{5}\binom{4}{i}\binom{5}{5-i}$$

In general:

> **Vandermonde's Theorem**
>
> Let n, m, k be integers with $n \geq 1$, $m \geq 1$, and $k \geq 0$. Then
> $$\binom{n+m}{k} = \sum_{i=0}^{k}\binom{n}{i}\binom{m}{k-i}$$

This can be proved using a *combinatorial proof*, a strategy that counts the same set two different ways and then equates the two counts. (A brief summation notation review is on page 230.)

Here is a summary of the counting formulas presented so far:

The Independent Tasks Principle

If a project can be decomposed into two independent tasks with n_1 ways to accomplish the first task and n_2 ways to accomplish the second task, then the project can be completed in $n_1 \cdot n_2$ ways.

The Mutually Exclusive Tasks Principle

If a project can be accomplished by completing exactly one of two mutually exclusive tasks, with n_1 ways to accomplish the first task and n_2 ways to accomplish the second task, then the project can be completed in $n_1 + n_2$ ways.

	Permutation With Order	**Combination** Without Order
Without Repetition	$P(n,k) = \frac{n!}{(n-k)!}$	$C(n,k) = \frac{n!}{k! \cdot (n-k)!}$
With Repetition	n^k	$C(n+k-1,k) = \frac{(n+k-1)!}{k! \cdot (n-1)!}$

Table 2: Arranging k elements from a set containing n distinct elements.

The Multinomial Counting Theorem

Suppose there exists a set of n items containing n_1 *identical* items of type 1, n_2 *identical* items of type 2, ..., and n_k *identical* items of type k, where $n = n_1 + n_2 + \cdots + n_k$. The number of visually distinguishable ways to arrange the n items in a row is

$$\text{multinomial}(n_1, n_2, \ldots, n_k) = \frac{n!}{n_1! \cdot n_2! \cdots n_k!}$$

Vandermonde's Theorem

Let n, m, k be integers with $n \geq 1$, $m \geq 1$, and $k \geq 0$. Then

$$\binom{n+m}{k} = \sum_{i=0}^{k} \binom{n}{i}\binom{m}{k-i}$$

127

Exercises

Solutions can be found at `http://www.mathcs.bethel.edu/~gossett/DMGN/` but solve them yourself before checking the answers. Before solving, identify the counting formula or formulas that you will use, and why they are appropriate for the exercise.

1. In how many visually distinguishable ways can 3 pennies, 2 nickels, 4 dimes, and 3 quarters be placed in a line?

2. The local pet shop has puppies, kittens, and hampsters. A mother of three children plans to let each child choose a pet. If she is only concerned with the kinds of animals (any two puppies are considered to be the same), and also is not concerned with which kind of pet a particular child chooses, in how many ways can the children choose pets. Once you have a count, list all the possible collections of pets (for example, ppp, ppk, pkh, etc.).

3. The game of dominoes uses a set of small rectangular tiles. Each tile is called a *domino*. A domino contains two equal-sized regions that contain zero or more dots, in fixed patterns. (A region with no dots is called a "blank.") The domino set has a maximum number of dots per region (usually 6, 9, or 12). Every pattern (number of dots) appears exactly one time with every other pattern on a domino. Dominoes with both regions containing the same pattern are called "doubles." The following diagram shows two dominoes: the 1–5 domino and a double 4.

   Create a formula $D(n)$ that computes the number of dominoes in a set whose highest tile is a double n. For example, $D(0) = 1$, since there is only one domino (double blanks). Also, $D(1) = 3$ since there will be three dominos: double blanks, blank–1, and double 1s.

4. Prove that the natural numbers can be placed in one-to-one correspondence with the non-negative rational numbers.

5. I have 10 pennies, 10 nickels, and 10 dimes. Suppose I put them all in a pile, close my eyes and randomly take 6 coins and determine how many cents in money I have. How many different amounts of money are possible? For extra credit, list all the different amounts that are possible.

6. Cynthia has 4 sheets of pink construction paper, 7 sheets of red construction paper, 2 sheets of yellow construction paper, and 3 sheets of white construction paper. She intends to make identical (except for color) valentines cards for 16 of her friends. In how many different ways can she hand out the cards? Could you answer this by listing all the different ways?

7. Miss Crabtree's first-grade class (22 students) plans to perform a play. She needs 1 student to be the scarecrow, 1 to be the farmer, 8 to be sheep, 4 to be crows, and 8 students to be corn stalks. In how many ways can she assign the parts?

8. Miss Crabtree wants her first-grade class (12 girls, 10 boys) to have a Thanksgiving party. She needs 5 students to help plan the party. Assume for this question that both the individual students chosen and the gender balance on the committee are of interest. In how many ways can the committee be formed? Solve this in two ways.

9. A business wants to identify all their products with a product code. The code will consist of 2 uppercase letters (from the set {A, B, C, D, E}), followed by 5 digits. The letters may not be repeated, but the digits may. How many distinct product codes exist?

10. Seven teams are participating in an invitational soccer tournament. Each team will bring 12 players. First, second, and third place will be determined. In addition, two players from each team will earn MVP (most valuable player) status. After the tournament, the 6 MVP players from the top three teams will get to attend a special banquet. Assuming any player might be designated MVP, in how many ways can the banquet guests be chosen?

Quiz Yourself

Without looking back at the material in *Counting 1* and *Counting 2*:

1. Write as many counting definitions as you can remember. Be as complete and precise as possible. Almost all of the words and symbols in the definitions are necessary.

2. Write as many counting theorems and principles as you can recall. Be complete and precise.

3. Make up an example for each counting theorem and principle. (You may look at the summaries on pages 107 and 127.)

Doing self-quizzes of this nature is a very effective way to thoroughly learn a subject. Initially, you may not do very well, but the effort you expend to retrieve the information helps to strengthen your brain's storage of that information. Here are two of the findings in *Make It Stick: the Science of Successful Learning*, by Brown, Rodediger, McDaniel.

> *Effortful retrieval* makes for stronger learning and retention. We're easily seduced into believing that learning is better when it's easier, but the research shows the opposite: when the mind has to work, learning sticks better. The greater the effort to retrieve learning, provided that you succeed, the more that learning is strengthened by retrieval. After an initial test, *delaying subsequent retrieval practice* is more potent for reinforcing retention than immediate practice, because delayed retrieval requires more effort.
> *Repeated retrieval* not only makes memories more durable but produces knowledge that can be retrieved more readily, in more varied settings, and applied to a wider variety of problems.

Notice the suggestion to leave some time between practice retrieval sessions. Trying to do it all the night before an exam is far less effective than numerous, shorter sessions scattered over many days prior to the exam.

Chapter 9

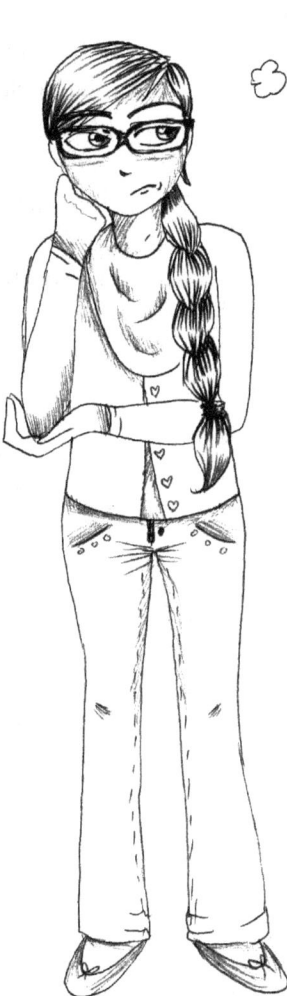

Isolde, before we start today, I have a question. You told me that the integers and the rational numbers are the same size. I trust you, so I believe it, but I don't have a clue as to why it is true.

I have seen a proof that the even integers and the set of all integers are sets of the same size (see page 120 for such a proof). The proof specifies a 1-1 correspondence between E and $\mathbb{N}$. But I don't see how to pair the integers and the rationals.

This is definitely not an idea that is intuitively true. The key feature is that both the integers and the rationals are infinite sets. Intuition is less reliable with infinite sets.

It is easy to see that the integers can't be any larger than the rationals. One proof that they are the same size shows that the rationals can not be larger than the integers.

To make things simpler, let's just show that the positive integers and the positive rationals are the same size. What I will actually do is to show how to pair a set that is even larger than the positive rationals with the positive integers.

More specifically, I will list all the positive rationals with lots of duplicates. For example, $\frac{2}{3}, \frac{4}{6}, \frac{6}{9}, \dots$ will each be listed, even though they all represent the same rational number.

I will proceed by listing this duplicated set of rationals in an infinite matrix. Each row will systematically list all the fractions with a given denominator. The columns will systematically list all the fractions with the same numerator.

$$\begin{array}{cccccc}
\frac{1}{1} & \frac{2}{1} & \frac{3}{1} & \frac{4}{1} & \frac{5}{1} & \frac{6}{1} \quad \cdots \\
\frac{1}{2} & \frac{2}{2} & \frac{3}{2} & \frac{4}{2} & \frac{5}{2} & \frac{6}{2} \quad \cdots \\
\frac{1}{3} & \frac{2}{3} & \frac{3}{3} & \frac{4}{3} & \frac{5}{3} & \frac{6}{3} \quad \cdots \\
\frac{1}{4} & \frac{2}{4} & \frac{3}{4} & \frac{4}{4} & \frac{5}{4} & \frac{6}{4} \quad \cdots \\
\frac{1}{5} & \frac{2}{5} & \frac{3}{5} & \frac{4}{5} & \frac{5}{5} & \frac{6}{5} \quad \cdots \\
\vdots & \vdots & \vdots & \vdots & \vdots
\end{array}$$

The clever idea is to match these entries with integers by going down diagonals.

$$\begin{array}{cccccc}
1 & 2 & 4 & \frac{4}{1} & \frac{5}{1} & \frac{6}{1} \quad \cdots \\
3 & 5 & \frac{3}{2} & \frac{4}{2} & \frac{5}{2} & \frac{6}{2} \quad \cdots \\
6 & \frac{2}{3} & \frac{3}{3} & \frac{4}{3} & \frac{5}{3} & \frac{6}{3} \quad \cdots \\
\frac{1}{4} & \frac{2}{4} & \frac{3}{4} & \frac{4}{4} & \frac{5}{4} & \frac{6}{4} \quad \cdots \\
\frac{1}{5} & \frac{2}{5} & \frac{3}{5} & \frac{4}{5} & \frac{5}{5} & \frac{6}{5} \quad \cdots \\
\vdots & \vdots & \vdots & \vdots & \vdots
\end{array}$$

This diagram shows that the positive rationals (which is not larger than the infinitely duplicated set of fractions in the matrix) cannot be larger than the set of positive integers. But we already know that the positive integers can't be larger than the positive rationals. Therefore, the two sets must be the same size.

$$|\mathbb{Z}| = |\mathbb{Q}|$$

That makes sense. So how do you prove that the set of real numbers is *larger* than the integers or the rationals? You certainly can't do it by finding a 1-1 correspondence.

$$|\mathbb{Z}| < |\mathbb{R}|$$

Georg Cantor, the mathematician who first formalized these kinds of ideas, *did* use a 1-1 correspondence in a proof.

His clever idea was to assume that there *was* a 1-1 correspondence, and then derive a contradiction.

The set of real numbers strictly between 0 and 1 is not larger than the entire set of real numbers. Cantor's idea will show that this set, denoted (0,1), is larger than the set of positive integers.

Let's assume that there is some way to pair up the numbers in (0,1) with the positive integers. The table of partners might look something like this. (Some entries may have infinite non-repeating digits.)

1	$0.3215\overline{0}$
2	$0.\overline{3}$
3	$0.\overline{142857}$
4	$0.12479312\cdots$
5	$0.444444111111\cdots$
6	$0.80\overline{0}$
$\vdots$	$\vdots$

Here is the scathingly brilliant part. Cantor created a real number, $x \in (0,1)$ that is *not* in the table. In fact, if we add x (by adjusting the integer assignments), then he would still be able to create *another* real number in (0,1) that is not in the table. That is the desired contradiction, because the table was assumed to contain *every* real number in (0,1).

Here is how to create x. First, no digit of x will be a 0 or a 9 (just to avoid issues such as $0.1\overline{9} = 0.2\overline{0}$). The first digit (after the decimal point) of x can be any digit between 1 and 8 (inclusive) that does not match the first digit after the decimal point of the first entry in the table. (So anything except a 3, 0, or 9 in the sample table.) In general, the nth digit after the decimal point must be a digit between 1 and 8 that does *not* equal the nth digit in row n of the table.

This contradiction was caused by the assumption that the real numbers in (0,1) could be placed in 1-1 correspondence with the positive integers. Since there will always be real numbers in (0,1) that can't be put in the table, (0,1) must be a larger set.

That is a really nice proof!

But that means that there is more than one size of infinity. What a curious idea!

Before we start today's session, I have one more thing to mention. My parents asked me to invite you to our house for dinner on either Friday or Saturday night—whichever works best for you.

I have already made a commitment for Friday night, but Saturday night would work. At what time should I show up? How formal is the event?

Saturday is great! That is the night that Logan can also come. (^_^)

Can you show up around 5:30? You don't need to bring anything. My mom said that poor, starving university students shouldn't be given additional expenses.

Although I would love to see you in formal evening wear, you just need to show up in your normal tutoring clothes.

Since that's settled, let's start today's session.

Algorithms

$\Theta(n)$

```
integer sequentialSearch (x, {a₀, a₁, a₂, ..., aₙ₋₁})
   for i = 0 to n − 1
      if x == aᵢ
         return i           # x == aᵢ so exit and return i
   return "not found"       # x did not match any of the as
end sequentialSearch
```

$\Theta(\log_2(n))$

```
integer binarySearch (x, {a₀, a₁, a₂, ..., aₙ₋₁})
   low = 0                  # 1st edge of the active portion
   high = n − 1             # 2nd edge of the active portion
   while low ≤ high
      mid = ⌊(low + high)/2⌋ # active portion midpoint (round down)
      if x > a_mid
         low = mid + 1       # ignore left half next iteration
      else if x < a_mid
         high = mid − 1      # ignore right half next iteration
      else
         return mid          # found x = a_mid at position mid
   return "not found"        # x is not in the list
end binarySearch
```

The topic for today's session is quite important for computer science majors. It's a mathematical tool for evaluating the efficiency of an algorithm.

An *algorithm* is an unambiguous set of steps for solving a problem in a finite amount of time. For example, the problem might be to sort a list of names into alphabetical order. There are numerous algorithms for doing this. Some are better than others (in a precise sense I will soon describe).

Today's Fight!
Algorithm A vs. Algorithm B

I know a bit about algorithms—Logan has been teaching me some computer programming.

That's great. I took an introductory programming course last semester. It was challenging – but actually pretty interesting. I am glad I enrolled.

The key idea is to decide which steps of the algorithm are the most time consuming. Then count how many times those steps need to be repeated.

There is one other factor that needs to be considered. The set of data with which the algorithm works will change. We need to do our analysis based on the size of the data set–not on a particular set of data. If the data set is small, pretty much any algorithm will be satisfactory. But as the data set increases in size, there can be dramatic differences in how long the algorithm takes to complete its task.

2 years
vs.
10 seconds

wow!

To see how dramatic this can be, consider the problem of sorting a list of names. Assume the list has 250 million names (fewer than the number of residents in America). One common group of sorting algorithms might take close to 2 years on a typical computer (assuming it had enough memory to store all the names). On the other hand, another group of sorting algorithms could do the task in under 10 seconds.

You mentioned groups of algorithms. How are the groups formed?

We first create, for each algorithm, a function that counts the number of time-consuming steps, as a function of the size, n, of the data set. Next, we group all the functions into similar groupings. For each grouping we elect a representative, called a *reference function*. For the sorting example I previously mentioned, the reference function for the slow group is $g_1(n) = n^2$ and for the faster group it is $g_2(n) = n \log_2(n)$. As n gets larger and larger, all the functions in the group have graphs that are almost identical to the graph of the reference function. However, the graphs for distinct reference functions are quite different.

Here is an example. One very simple algorithm for sorting is named *selection sort*. The idea is to scan the list and find the name that is the first in alphabetical order. Swap it with the current first name in the list. Now focus on the sublist that omits the first name (since that name is now in proper position). Keep doing this on decreasing sublists until the list has only one element, then quit. On every scan, you need to compare each new item to the currently found alphabetically smallest name in the sublist. The total number of comparisons is $\frac{1}{2}n^2 + \frac{1}{2}n$. Looking at the graphs of $f(n) = \frac{1}{2}n^2 + \frac{1}{2}n$ and $g_1(n) = n^2$, we see that as n gets big, the two graphs have a similar shape, especially when compared to a function like $h(n) = n^3$ or $k(n) = n$.

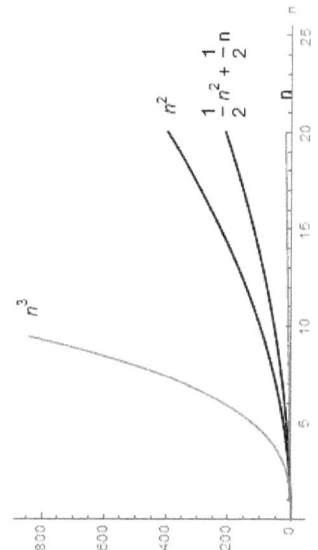

To make this more precise, we say that $f(n)$ is in the same group as $g(n)$ as long as we can find constants c_1 and c_2 so that, for large enough values of n, the graph of $f(n)$ is always between the graphs of $c_1 \cdot g(n)$ and $c_2 \cdot g(n)$. For the selection sort, we can use $c_1 = \frac{1}{4}$ and $c_2 = 1$ (see the graph). There are many other choices for c_1 and c_2 that would make the inequality true.

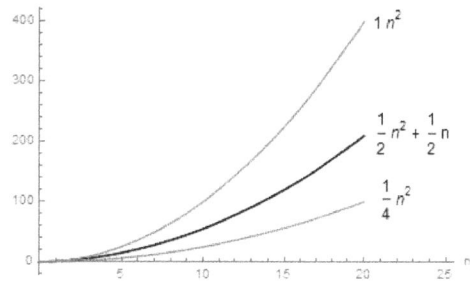

$$\frac{1}{4}n^2 \leq \left(\frac{1}{2}n^2 + \frac{1}{2}n\right) \leq n^2$$

So the idea is to group the functions that sort of look like each other?

Yes. The "sort of look like each other" is captured by the requirement that $f(n)$ is eventually trapped between fixed multiples of $g(n)$. Here is another example. Let $f(n) = 3n^3 + 4n^2 - 5n + 50$ and $g(n) = n^3$. Then we can pick $c_1 = 2$ and $c_2 = 4$. As long as $n \geq 5$ it is always true that $2g(n) \leq f(n) \leq 4g(n)$.

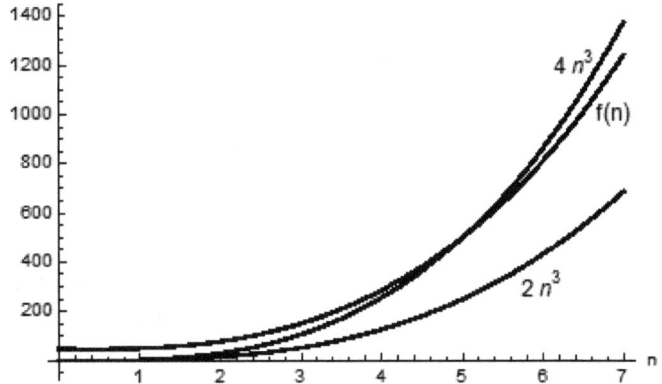

This can all be made quite formal with a few scary looking definitions. However, the definitions accurately capture the intuitive ideas I just presented. The first one says that $f(n)$ should eventually be smaller than a multiple of $g(n)$. If that is true, then f is in the "grows no faster than g" club of functions. We denote this by saying f is in the set Big-$\mathcal{O}$ of g.

Definition *Big-$\mathcal{O}$*

The function f is said to be in big-$\mathcal{O}$ of g, pronounced "big oh" and denoted $f \in \mathcal{O}(g)$, if there are positive constants, c_2 and n_0, such that, for all $n \geq n_0$, $|f(n)| \leq c_2|g(n)|$.

Another way to express this is $f \in \mathcal{O}(g)$ if and only if

$$\exists c_2 \in \mathbb{R}^+, \exists n_0 \in \mathbb{R}^+, \forall n \in \mathbb{R}^+, [(n \geq n_0) \to (|f(n)| \leq c_2|g(n)|)]$$

$\mathbb{R}^+$ is the set of nonnegative real numbers. This definition takes us part of the way towards the idea we want, but it is insufficient on its own. The reason is that, for example, $n \in \mathcal{O}(n^2)$, but n^2 grows much faster than does n. We would not want to declare that "n sort of looks like n^2." The next definition fixes this deficiency.

138

Definition *Big-Ω*

The function f is said to be in big-Ω of g, pronounced "big omega" and denoted $f \in \Omega(g)$, if there are positive constants, c_1 and n_0, such that, for all $n \geq n_0$, $|f(n)| \geq c_1|g(n)|$. Another way to express this is $f \in \Omega(g)$ if and only if

$$\exists c_1 \in \mathbb{R}^+, \exists n_0 \in \mathbb{R}^+, \forall n \in \mathbb{R}^+, [(n \geq n_0) \rightarrow (|f(n)| \geq c_1|g(n)|)]$$

This definition says that f is in the "grows faster than g" club of functions. If we combine these two definitions, we get the definition for which we are really looking.

Definition *Big-Θ*

The function f is said to be in big-Θ of g, pronounced "big theta" and denoted $f \in \Theta(g)$, if $f \in \mathcal{O}(g) \cap \Omega(g)$.

These definitions are a bit tedious to use in practice, but there is some freedom. For a given pair of functions, f and g, for which $f \in \Theta(g)$, there are infinitely many choices of n_0 and c_1, c_2 that will satisfy the definitions, due to the inequalities in the definitions.

Fortunately, there are several theorems that provide useful shortcuts in identifying a good reference function, g, for a function, f, that counts the number of time-consuming steps in an algorithm. I will mention just one of them.

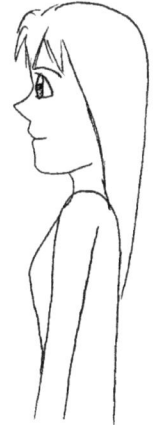

Theorem *Big-Θ and Polynomials*

Let f be the polynomial function
$f(n) = a_k n^k + a_{k-1} n^{k-1} + \cdots + a_1 n + a_0$,
where $a_k \neq 0$. Then $f \in \Theta(n^k)$.

So, using this theorem, what would be the proper Big-Θ reference function for an algorithm with $f(n) = 5n^4 + 7n^3 + 9$?

$f \in \Theta(5n^4)$?

Almost. Look at the theorem again. The convention is to drop the leading coefficient, a_k. The correct answer is $f \in \Theta(n^4)$.

As n gets bigger and bigger, the terms of degree less than 4 contribute relatively little to the value of the function. That is, the term $5n^4$ grows much faster than the term $7n^3 + 9$. The function grows very much like $g(n) = n^4$.

139

I have a question. In the definitions of big-$\mathcal{O}$ and big-Ω, I am not really clear on the role of n_0.

Look back (on page 138) at the example with $g(n) = n^3$ and $f(n) = 3n^3 + 4n^2 - 5n + 50$. When $n < 5$, $f(n) > 4g(n)$. But we needed $f(n) < 4g(n)$ if we choose $c_2 = 4$. Changing c_2 won't eliminate the problem—there will be another region (instead of $n < 5$) where f is larger than $c_2 g$.

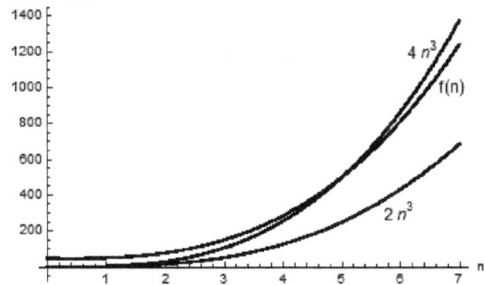

The number n_0 expresses the idea that we only care about algorithm performance for large data sets. So we only need to have $|f(n)| \leq c_2|g(n)|$ for large enough values of n ($n \geq n_0$). It is ok to have the inequality reversed when n is small.

That makes sense. Thanks.

It is helpful to have standard reference functions to provide a sense of how various algorithms compare to each other. The following list contains some of the most common reference functions, ranked by their complexity.

The higher a reference function is in the list, the better it performs. Cubic functions can be pretty slow, but still useful. An exponential function is generally not useful – it may do ok on small data sets, but even for medium-sized data sets it may take too long to be practical.

Sometimes the problem that we are solving imposes a minimum complexity on any correct algorithm. In linear algebra, there is no way to multiply two matrices with less than a $\Theta(n^3)$ algorithm (unless the matrices have some special properties that we can exploit).

Category	Reference Function	Complexity Rank
Logarithmic	$\log_2(n)$	Excellent
Linear	n	Very Good
$n \log n$	$n \log_2(n)$	Good
Quadratic	n^2	Fair
Cubic	n^3	Slow
Exponential	a^n for $a > 1$	Worst (Useless)

I will conclude today's session with a comparison of two algorithms which solve the same problem: searching for an item in a list. The algorithm also needs to work properly if the item is not in the list.

Lily, if I were to hand you a randomly ordered list of names and asked you to find *Demelza Poldark*, how would you complete the task?

Since the list is randomly ordered, I would just start at the top and read down the list until I found the name I was searching for.

I can't think of a better approach, unless it is to close my eyes and point at random spots in the list and hope I hit the correct name. But I could keep doing that for a long time and never hit the right spot.

You are correct. The best we can do with a randomly ordered list is to systematically examine each item in the list. Your approach is called a *sequential search*. If you understand *pseudocode*, the algorithm can be written as follows (you can ignore this if you are not familiar with pseudocode or programming).

In the first line, x is the item we are looking for and $\{a_0, a_1, a_2, \ldots, a_{n-1}\}$ is the list of all items. (Computer scientists tend to start counting at position 0.)

```
integer sequentialSearch (x, {a₀, a₁, a₂, ..., aₙ₋₁})
    for i = 0 to n − 1
        if x == aᵢ
            return i          # x == aᵢ so exit and return i
    return "not found"        # x did not match any of the as
end sequentialSearch
```

In the worst case, we will find the item we are looking for in the last place (or not find it at all), taking n comparisons. If we keep rearranging the random order of the items, on average we would expect to find the item about halfway through the list, taking $\frac{n}{2}$ comparisons. In either case, the algorithm is in $\Theta(n)$.

So it is pretty efficient (rated very good in the complexity rankings).

Yes, it *is* pretty good—as good as possible if the list is randomly ordered. But what if the list is in alphabetical order? When your parents were young, people used phone books—lists of phone numbers sorted by the name of the person who was assigned the number.

So how would you go about finding the name if the list is in alphabetical order?

I could make a guess as to where in the list the name would be, then look at that area. Since the "P" in "Poldark" is a bit past the middle of the alphabet, I would first look a bit past the middle of the list. I would then look at the name at the position I guessed and either look before or after that position depending on how it compares to "Poldark".

That is a great approach! It is similar in spirit to the next algorithm: *binary search*. In this algorithm, we look first at the middle location in the list and compare that item with the item we are searching for. We keep repeating until we either find the item, or run out of list to search. The pseudocode is next.

```
integer binarySearch (x, {a₀, a₁, a₂, ..., aₙ₋₁})
    low = 0                # 1st edge of the active portion
    high = n − 1           # 2nd edge of the active portion
    while low ≤ high
        mid = ⌊(low + high)/2⌋ # active portion midpoint (round down)
        if x > a_mid
            low = mid + 1  # ignore left half next iteration
        else if x < a_mid
            high = mid - 1 # ignore right half next iteration
        else
            return mid     # found x = a_mid at position mid
    return "not found"     # x is not in the list
end binarySearch
```

Consider a worst case search. At each step, the list is cut in half. How many times can that be done? Suppose the list has $n = 2^k$ items. We want to keep dividing the list in half until there is only one item left. There will be k possible divisions by 2, where $k = \log_2(n)$. (See page 229.)

Even when $n \neq 2^k$, both the average and worst case complexity of binary search is in $\Theta(\log_2(n))$. There are two reasons as to why it can do better than sequential search: (1) it can exploit the sorted nature of the list (and we did not count the work done to sort the list), and (2) it does not need to examine every item in the list.

In the worst case, to find an item in a list of $n = 10,000$ items, the algorithm above would take 40 comparisons ($3\lfloor \log_2(n) \rfloor + 1$), instead of 10,000 for sequential search. (I have not fully justified these claims.)

Definitions

Algorithm An *algorithm* is an unambiguous set of steps for solving a problem in a finite amount of time.

Big-$\mathcal{O}$ The function f is said to be in big-$\mathcal{O}$ of g, pronounced "big oh" and denoted $f \in \mathcal{O}(g)$, if there are positive constants, c and n_0, such that, for all $n \geq n_0$, $|f(n)| \leq c|g(n)|$.

Another way to express this is $f \in \mathcal{O}(g)$ if and only if

$$\exists c \in \mathbb{R}^+, \exists n_0 \in \mathbb{R}^+, \forall n \in \mathbb{R}^+, [(n \geq n_0) \rightarrow (|f(n)| \leq c|g(n)|)]$$

Big-Ω The function f is said to be in big-Ω of g, pronounced "big omega" and denoted $f \in \Omega(g)$, if there are positive constants, c and n_0, such that, for all $n \geq n_0$, $|f(n)| \geq c|g(n)|$. Another way to express this is $f \in \Omega(g)$ if and only if

$$\exists c \in \mathbb{R}^+, \exists n_0 \in \mathbb{R}^+, \forall n \in \mathbb{R}^+, [(n \geq n_0) \rightarrow (|f(n)| \geq c|g(n)|)]$$

Big-Θ The function f is said to be in big-Θ of g, pronounced "big theta" and denoted $f \in \Theta(g)$, if $f \in \mathcal{O}(g) \cap \Omega(g)$.

Theorems

Big-Θ and Polynomials Let f be the polynomial function
$f(n) = a_k n^k + a_{k-1} n^{k-1} + \cdots + a_1 n + a_0$,
where $a_k \neq 0$. Then $f \in \Theta(n^k)$.

Some Common Reference Functions

Category	Reference Function	Complexity Rank
Logarithmic	$\log_2(n)$	Excellent
Linear	n	Very Good
$n \log n$	$n \log_2(n)$	Good
Quadratic	n^2	Fair
Cubic	n^3	Slow
Exponential	a^n for $a > 1$	Worst (Useless)

Exercises

Solutions can be found at `http://www.mathcs.bethel.edu/~gossett/DMGN/`.

1. For each of the following function pairs, use the formal definitions to show that $f \in \Theta(g)$. Do this by finding suitable values for n_0, c_1, c_2.

   (a) $f(n) = 3n + 40 \qquad g(n) = n$

   (b) $f(n) = 3n^3 + 17n^2 + 4 \qquad g(n) = n^3$

   (c) $f(n) = 6n \log_2(n) + 3n \qquad g(n) = n \log_2(n)$

   (d) $f(n) = (3n + 2)(2 \log_2(n) + n) \qquad g(n) = n^2$

2. For each of the following functions, use the Big-Θ and Polynomials theorem to find a suitable reference function.

   (a) $f(n) = 17n + 45$

   (b) $f(n) = \frac{1}{3}n^2 + 6n - 9$

   (c) $f(n) = 4$

3. Suppose a computer is capable of executing one time consuming step in 10^{-9} seconds. Suppose that four algorithms are available, each with a different big-Θ reference function. Assuming the data set has 5,000,000 items, fill in the following chart to show how long each algorithm will take to complete. Convert your answer into hours, minutes, and seconds.

Reference Function	n^2	$n \log_2(n)$	n	$\log_2(n)$
Time				

4. Suppose a computer is capable of executing one time consuming step in 10^{-9} seconds. Suppose that four algorithms are available, each with a different big-Θ reference function. Assuming the algorithm can run for 1 second , how large a data set can each algorithm complete? (You will need to use a numeric approximation for the $n \log_2(n)$ reference function.)

Reference Function	n^3	n^2	$n \log_2(n)$	n
Size of Data Set				

144

Chapter 10

Isolde! I am so happy that you were able to come to my house for dinner. My parents were very impressed.

Thanks for inviting me. I had a great time and I really liked your family.

Does that mean you like Logan? I am pretty sure that he likes you. If you two got married we could be sisters-in-law!

I knew this was coming sometime ...

Yes Lily, it would be nice if you and I were to be sisters-in-law. I like you very much, and I enjoy being with Logan. However, it is much too soon for me to give any serious thought to marriage. I still need to finish all my classes and then student teach. After that I need to find a job, possibly move to a new city, and then get settled into my new job. All of this would be much harder for me if I were also throwing romance and marriage into the mix.

Cupid bites the dust!

That does make sense. I'm sorry if I embarrassed you or sounded too pushy. Sometimes I get a bit carried away with my dreams.

Let me apologize with a new puzzle. A long time ago there was a mathematician named Martin Gardner who wrote a regular puzzle column for *Scientific American*. My dad bought a book that gathered a number of those puzzles together. I found a very fun one in that book.

Here is the puzzle. Start with a chess board and 32 dominoes. Assume each domino perfectly covers two squares on the chess board. It is easy to cover all 64 squares using the 32 dominoes. But, what if you cut off two corners. Can you use 31 dominoes to cover all the squares? (Try to solve this on your own before looking at the next page.)

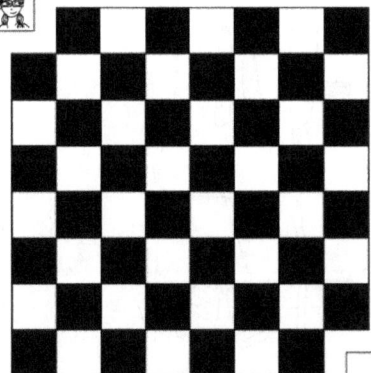

146

Hmm ... If it were the top two corners missing, it would be easy to cover with dominoes.

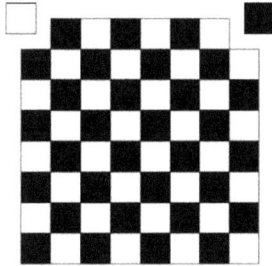

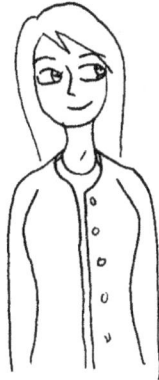

Perhaps I can consider smaller chessboards. If I start with a 2 by 2 board and remove the corners, it is clear that I can't use a domino to cover the remaining squares.

A 3 by 3 chessboard won't work because, after removing the two corners, I would have an odd number of squares left.

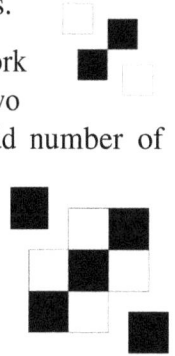

Let me consider a 4 by 4 chessboard. It will leave an even number of squares.

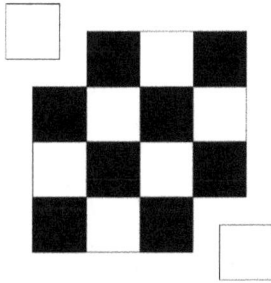

Isolde tries several ways to cover the squares with dominoes.

Well, I am not getting it yet. At least the 4 by 4 chessboard is a bit easier to work with, so I think I will stick with it for a bit. I think I will need some dominoes to be horizontal and some to be vertical.

muahaha

Several minutes later ...

I think I am stuck, Lily. Can I get a hint?

Think about how a chessboard is structured. In particular, think about the colors.

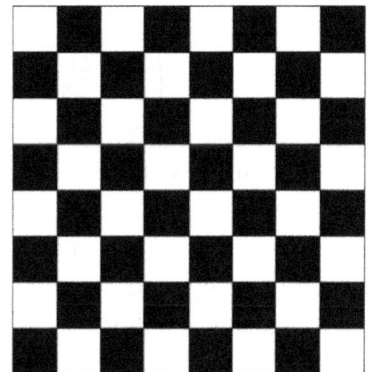

Try the hint on your own before turning the page.

Well . . . , the colors alternate.

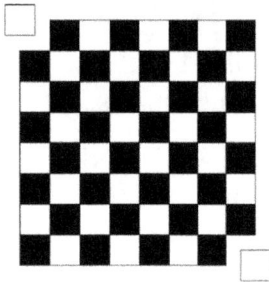

And a domino covers two adjacent squares, so it must always cover two different colors.

But this means that I can't cover the chessboard with the missing corners because they both have the same color. After the first 30 dominoes all the remaining white squares will be covered, yet two black squares will still be without a covering domino. But black squares are never adjacent. So the task is impossible to complete.

Isn't that a great puzzle!

Yes, it is. It has a very simple solution, but that simple, very clever solution is not easy to find.

Should we move on to today's lesson?

I have a question first. Earlier, you mentioned student teaching. When does that happen?

I student teach next semester. That means I won't have time to tutor you after this semester is over.

I will miss our weekly sessions!

Do you think you could do your student teaching at my school? It would be great to have you as my teacher. Some of my friends at school would also like to be in your class.

I have little control over which school I end up in. The University's Education Department and the local school districts negotiate the assignments. I would love to be in your school, but it is not likely.

Recursion

Today's topics are recursion and recurrence relations. They are, in spirit, the same idea in two different contexts.

Recursion is a programming technique that completes a task by using an algorithm that in turn uses one or more copies of itself with a smaller or simpler data set.

A *recurrence relation* is a formula that defines numbers in a sequence in terms of previous values in the sequence.

We will look at examples of each today.

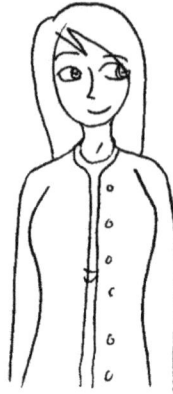

Will you help me complete this task?

I will ask my own reflection to help.

Let me illustrate the recursive approach to creating an algorithm by considering the gcd of two integers. Recall the definition.

Definition *Greatest Common Divisor*

Let a and b be integers that are not both 0. The *greatest common divisor* (gcd) of a and b is a positive integer d such that

- $d \mid a$ and $d \mid b$.
- If c divides both a and b, then $c \mid d$.

The greatest common divisor of a and b is denoted by $\gcd(a, b)$.

The definition makes it clear that $\gcd(a, b) = \gcd(b, a)$. I will prove that whenever $a \leq b$, then $\gcd(a, b) = \gcd(a, b - a)$.

Theorem

$\gcd(a, b) = \gcd(a, b - a)$

Here is the proof. First assume that $d = \gcd(a, b)$. Then there exist integers, n and m with $a = dn$ and $b = dm$. Therefore $b - a = dm - dn = d(m - n)$, which implies that d is a common divisor of a and $b - a$.

Now assume that c is any common divisor of a and $b - a$. Then there exist integers s and t such that $a = cs$ and $b - a = ct$. But this implies that $b = (b - a) + a = ct + cs = c(t + s)$, so c is a common divisor of a and b. But d is the greatest common divisor of a and b so $c \mid d$ must be true.

Since $d \mid a$ and $d \mid (b - a)$ and also if $c \mid a$ and $c \mid (b - a)$, then $c \mid d$, the definition asserts that d is the gcd of a and $b - a$. $\square$

Here is how we can use this theorem to create a recursive algorithm to find the gcd. Let's assume we want gcd(12, 28).

I don't know gcd(12, 28), but I know it is the same as gcd(12, 28 − 12) = gcd(12, 16). Notice that gcd(12, 16) is a smaller gcd problem (one parameter is the same and the other is smaller than that in the original problem). Suppose I don't know the value of gcd(12, 16). In that case, I can again use recursion to find gcd(12, 16 − 12) = gcd(12, 4). But gcd(12, 4) = 4, so I am done: gcd(12, 28) = 4.

$$\text{gcd}(12,28)$$
$$\downarrow$$
$$\text{gcd}(12,16)$$
$$\downarrow$$
$$\text{gcd}(12,4)$$
$$\downarrow$$
$$4$$

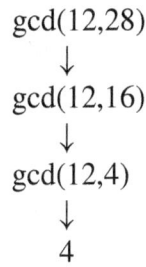

So recursion is replacing a problem with a smaller version of the same kind of problem? How do you know when to stop finding smaller problems?

You have raised two important issues with recursion:

1. The *first* thing you need to do in any recursive algorithm is to check if you are done yet; are you at a subproblem for which the solution is obvious? (You need one or more base cases.)

2. You need a strategy for putting the solutions to the smaller problems together in a manner that correctly solves the original problem.

In my gcd example, the strategy for putting the subproblem solutions together was easy; the subproblem and the original problem have the same solution.

Here is some pseudocode for the gcd algorithm.

```
integer gcd (integer a, integer b)
  if (a > b)
    swap a and b
  if a == 0
    return b
  if a == 1
    return 1
  return gcd(a, b - a)
```

The algorithm can stop at gcd(0, b) = b or at gcd(1, b) = 1. Even better algorithms are possible. For example, the algorithm could use the theorem
$$\text{gcd}(a, b) = \text{gcd}(a, b \bmod a).$$

Recursive algorithms are more at home in a computer science course on programming, so I won't say any more on that topic (although some very important ideas would still need to be presented if you want to master recursion).

I will introduce recurrence relations by using a famous sequence of numbers that was introduced in Europe by Leonardo Bonacci (known as Fibonacci) around 1200 AD, but dating back to perhaps around 700 AD in India.

The sequence starts as
$$1, 1, 2, 3, 5, 8, 13, \ldots$$

Oh! I know this! Each number in the sequence is the sum of the two previous entries. So the entry after 13 will be $8 + 13 = 21$.

That is right. (Except that the first two entries are not sums.) If we denote the entries as $f_0, f_1, f_2, \ldots$, then

$$f_0 = 1$$
$$f_1 = 1$$
$$f_n = f_{n-1} + f_{n-2} \quad \text{for } n \geq 2$$

Ok. I see how each entry (after the first two) is defined in terms of earlier entries. So what more is involved with this topic?

Recurrence relations are useful for solving certain kinds of problems. For these problems, it might be relatively easy to specify a solution by using a recurrence relation. But that is not where we typically stop. What we usually want is a nice closed-form formula. That is, a nice formula that expresses f_n as a function of n (that doesn't involve any other f_k).

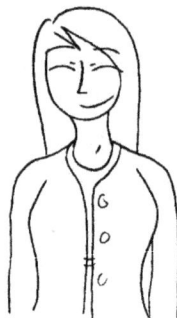

Here is a question Lily: what is f_{100}?

I suppose I could find it if I had enough time and a good calculator.

Actually, your calculator might let you down. I used some software to determine that
$f_{100} = 54{,}224{,}848{,}179{,}261{,}915{,}075.$

There are two phases for using recurrence relations: (1) find a solution for the problem that can be expressed using a recurrence relation and (2) solve the recurrence relation to produce a closed-form formula for the nth term in the sequence.

It is not trivial to carry out step (2) for the Fibonacci sequence, but it turns out that for $n \geq 0$

$$f_n = \left(\frac{5 + \sqrt{5}}{10}\right)\left(\frac{1 + \sqrt{5}}{2}\right)^n$$
$$+ \left(\frac{5 - \sqrt{5}}{10}\right)\left(\frac{1 - \sqrt{5}}{2}\right)^n$$

I will illustrate this two-phased process with an example. Suppose that I am constructing a quiz. I have two kinds of problems that can be on the quiz. The first kind are **Single**-part questions that should take 1 minute to answer. The second kind are **Two**-part questions that take 2 minutes. These have a and b parts, but the parts might appear in either order. Denote the two orders by T_{ab} and T_{ba}. Here are some examples.

1. Find $\gcd(27, 72)$.
2. Let $x = 7$ and $y = 212$.
   (a) Does $x \mid y$?
   (b) Is $212 \equiv 36 \bmod x$?

Question 2 could have the contents of the (a) and (b) parts reversed. ("Does $x \mid y$" could be part (b).)

Suppose I am only interested in the possible patterns for a quiz that lasts n minutes. For example, with a 2-minute quiz I have 3 possible patterns: SS, T_{ab}, T_{ba}. How many patterns are there for an n-minute quiz? Call the answer a_n.

As a start, Lily, how many patterns are there for 1-minute and 3-minute quizzes?

Well, a_1 is easy, since there is only time for a single-part question. So $a_1 = 1$. I can just list all the ways to get a 3-minute quiz: $SSS, ST_{ab}, ST_{ba}, T_{ab}S, T_{ba}S$. So $a_3 = 5$.

Do you think you could create a recurrence relation that describes how you could construct a_n in terms of a_{n-1} and a_{n-2}? Try building a_4 from a_3 and a_2 to help guide your intuition.

There are three patterns for a quiz that lasts for 2 minutes: SS, T_{ab}, T_{ba} and five patterns for one that lasts for 3 minutes: $SSS, ST_{ab}, ST_{ba}, T_{ab}S, T_{ba}S$. So $a_3 = 5$.

I can turn a 3-minute pattern into a 4-minute pattern by appending a single-part question. (3-minute quiz $\rightarrow$ 3-minute quiz S) That would contribute a total of $a_3 = 5$ patterns to a_4: $SSSS, ST_{ab}S, ST_{ba}S, T_{ab}SS, T_{ba}SS$.

3-minute quiz $\rightarrow$ 3-minute quiz S
2-minute quiz $\rightarrow$ 2-minute quiz SS
2-minute quiz $\rightarrow$ 2-minute quiz T_{ab}
2-minute quiz $\rightarrow$ 2-minute quiz T_{ba}

There are three ways to extend a 2-minute quiz into a 4-minute quiz: append SS, or T_{ab}, or T_{ba}. So that will contribute three 4-minute patterns for each 2-minute pattern. I suppose then that $a_n = a_{n-1} + 3a_{n-2}$.

That would make $a_4 = a_3 + 3a_2 = 5 + 3 \cdot 3 = 14$.

That is close, but there is a subtle error.

In order to see the error, let's build a_3 from a_2 and a_1. Informally, a_2 contributes one pattern to a_3 for each distinct pattern in a_2: SS, T_{ab}, T_{ba} become SSS, $T_{ab}S$, $T_{ba}S$. So for this part you were correct.

There is only one pattern (S) for a 1-minute quiz, so a_1 contributes the following extensions into 3-minute quizzes: $S\underline{SS}$, $S\underline{T_{ab}}$, $S\underline{T_{ba}}$.

There are 6 patterns, just as your recurrence relation predicts. But there is an error. Do you see what went wrong, Lily?

from a_2:
   SSS, $T_{ab}S$, $T_{ba}S$
from a_1:
   SSS, ST_{ab}, ST_{ba}

Hmm .... Oh! I see! The pattern SSS shows up twice, so it was double-counted.

Yes. In fact, we can always get to a pattern of the form $\#\#\#SS$ on an n-minute quiz in two ways:

- Start with a pattern for an $(n-1)$-minute quiz (but that ends with S) and then add another S.

- Start with a pattern for an $(n-2)$-minute quiz and append SS.

In effect, the second option is redundant. So there are really only 2 *new* ways to build patterns of time-size n from patterns of time-size $n-2$. The correct recurrence relation is therefore: $a_n = a_{n-1} + 2a_{n-2}$.

A pattern of length n will end with S, T_{ab}, or T_{ba}. Removing the last symbol leads to a pattern of length $n-1$ or $n-2$. So I can build any pattern of length n by appending either T_{ab} or T_{ba} to a pattern of length $n-2$, or appending S to a pattern of length $n-1$.

Correct. There is one more issue to discuss before I move over to phase 2 (where we convert the recurrence relation into a closed-form solution).

Suppose I want to also consider 0-minute quizzes. What should be the proper value for a_0? It is tempting to say $a_0 = 0$; should a no-question-quiz really count as a quiz? But there is a better interpretation.

We declare that there is one way to create a 0-minute quiz – namely, do nothing. There is a compelling justification for this interpretation: $a_0 = 1$ is the value that makes a_0 work properly with the recurrence relation we just derived. Here is the proper way to write the recurrence relation:

$$a_0 = 1$$
$$a_1 = 1;$$
$$a_n = a_{n-1} + 2a_{n-2} \text{ for } n \geq 2$$

Now that we know we have a correct recurrence relation, it is time to find a closed-form solution. The advantage is that if we want the value of a_{1000}, we won't need to calculate all of the prior values, as would be necessary using only the recurrence relation.

There are a number of techniques available for this phase, including back substitution and generating functions. I will use a technique that works well for a certain class of recurrence relations.

So ..., not all recurrence relations are created equal?

You are correct. The special class has three properties that are captured in the next definition.

Definition *Linear; Constant Coefficients; Homogeneous*

A recurrence relation for the sequence $\{a_n\}$ is a *linear recurrence relation with constant coefficients* if it is in the form

$$a_n = c_1 a_{n-1} + c_2 a_{n-2} + \cdots + c_k a_{n-k} + f(n)$$

for some constants $c_1, c_2, \ldots, c_k$.
The recurrence relation is a *linear homogeneous recurrence relation with constant coefficients* if $f(n) = 0$ for all n.

I will abbreviate the phrase "linear homogeneous recurrence relation with constant coefficients" as LHRRWCC.

Here are some examples:

$a_n = 4a_{n-1} + 6a_{n-2} - 2a_{n-3}$ LHRRWCC

$b_n = 3b_{n-1}^2 + 4b_{n-2}$ b_{n-1}^2 is not linear

$c_n = n^2 c_{n-1} + 3c_{n-2}$ the coefficient n^2 is not constant

$d_n = 4d_{n-1} + 9d_{n-2} + 7$ $+7$ is not homogeneous

The conversion to closed-form requires two steps:

1. Find the characteristic equation for the recurrence relation and finds its roots. Use these roots to create the general closed-form solution.

2. Use the base values of the recurrence relation and the general solution to form a system of linear equations. The solutions to this system are the coefficients for the desired specific closed-form solution to the recurrence relation.

Step 1 involves the characteristic equation.

Definition *The Characteristic Equation*

The *characteristic equation* of the recurrence relation
$$a_n = c_1 a_{n-1} + c_2 a_{n-2} + \cdots + c_k a_{n-k}$$
is
$$x^k - c_1 x^{k-1} - c_2 x^{k-2} - \cdots - c_{k-1} x - c_k = 0$$

This is easy: just count the number of terms on the right (k), then convert a_n into x^k. Now move the terms on the right over to the left. The term a_{n-i} becomes x^{k-i}.

It is *really* important that the recurrence relation has been written with the subscripts on the a_i terms appearing in *decreasing order*.

So, Lily, can you identify the characteristic equation for our example: $a_n = a_{n-1} + 2a_{n-2}$? Also, can you find the roots?

Would it be $x^2 - x - 2 = 0$? I could use the quadratic formula to find the roots, but this equation is easy to factor: $x^2 - x - 2 = (x-2)(x+1) = 0$. Thus, $x = 2, -1$ are the roots.

That is correct. And we got lucky: the two roots are distinct. If the characteristic equation has a repeated root, what comes next would be a bit more complicated.

Did we really get lucky, or did you design the example to have distinct roots?

No comment.

Here is the general closed-form solution:

$$a_n = \theta_1 2^n + \theta_2 (-1)^n$$

where θ_1 and θ_2 are constants we still need to calculate.

There is a theorem which asserts that no matter what we choose for θ_1 and θ_2, this expression will satisfy the original recurrence relation. However, unless we choose the proper values for θ_1 and θ_2, $a_0 = 1$ and $a_1 = 1$ will *not* be true.

Step 2 uses our knowledge of the two base values: $a_0 = 1$ and $a_1 = 1$. Substitute $n = 0$ and then $n = 1$ into the general solution.

$$1 = a_0 = \theta_1 2^0 + \theta_2 (-1)^0$$
$$1 = a_1 = \theta_1 2^1 + \theta_2 (-1)^1$$

The system of equations can be rewritten as:

$$\theta_1 2^0 + \theta_2(-1)^0 = 1$$
$$\theta_1 2^1 + \theta_2(-1)^1 = 1$$

or as

$$\theta_1 + \theta_2 = 1$$
$$2\theta_1 - \theta_2 = 1$$

Lily, can you solve this?

Sure. From the second equation, I know that $\theta_2 = 2\theta_1 - 1$. Substitution into the first equation leads to $\theta_1 + (2\theta_1 - 1) = 1$ or $3\theta_1 = 2$. So $\theta_1 = \frac{2}{3}$ and then $\theta_2 = \frac{1}{3}$.

So the final closed-form solution is

$$a_n = \frac{2}{3} \cdot 2^n + \frac{1}{3} \cdot (-1)^n$$

Does this work? For $n = 2$ it produces

$$a_2 = \frac{2}{3} \cdot 2^2 + \frac{1}{3} \cdot (-1)^2 = \frac{2}{3} \cdot 4 + \frac{1}{3} = 3$$

So the closed-form solution works for $n = 2$. It is easy to make sure we didn't make a mistake somewhere in the process. Just make a table of values, with one column calculated from the recurrence relation and the next column calculated from the closed-form solution. If they are the same, we can be pretty sure the solution is correct.

Here is the table. The two base conditions are used for the first two entries in the middle column.

n	$a_n = a_{n-1} + 2a_{n-2}$	$a_n = \frac{2}{3} \cdot 2^n + \frac{1}{3} \cdot (-1)^n$
0	1	1
1	1	1
2	3	3
3	5	5
4	11	11
5	21	21

I would like to buy a 10-minute quiz.

No problem. There are 683 from which to choose.

There is one final point I want to emphasize. The proper way to express the closed-form solution is:

$$a_n = \frac{2}{3} \cdot 2^n + \frac{1}{3} \cdot (-1)^n \quad \text{for } n \geq 0$$

I have added the phrase "for $n \geq 0$". The recurrence relation is valid for $n \geq 2$, but the closed-form works for the base values as well.

$$a_0 = 1$$
$$a_1 = 1;$$
$$a_n = a_{n-1} + 2a_{n-2} \quad \text{for } n \geq 2$$

That is all pretty cool, but I need time to think about it.

No problem. That is all for today.

Definitions

Recursive Algorithm *Recursion* is a programming technique that completes a task by using an algorithm that in turn uses one or more copies of itself with a smaller or simpler data set.

- The *first* thing you need to do in any recursive algorithm is to check if you are done yet. You need one or more base cases.

- You also need a strategy for putting the solutions to the smaller problems together in a manner that correctly solves the original problem.

Recurrence Relation A *recurrence relation* is a formula that defines numbers in a sequence in terms of previous values in the sequence.

- To properly specify a recurrence relation for a_n, you need as many base cases as there are terms of the form a_{n-i} on the right.

- You also need to add the phrase "for $n \geq k$", where there are k terms of the form a_{n-i} on the right. (This assumes you start with a_0.)

Closed-form Formula Suppose a sequence, $\{a_n\}$, has been specified by a recurrence relation. A *closed-form formula* is a formula that expresses the value of a_n directly as a function of n, without using any subexpressions of the form a_{n-k}.

- The proper way to express a closed-form formula for a_n includes a phrase of the form "for $n \geq 0$" (assuming the sequence starts at a_0).

LHRRWCC A recurrence relation for the sequence $\{a_n\}$ is a *linear homogeneous recurrence relation with constant coefficients* if it is in the form

$$a_n = c_1 a_{n-1} + c_2 a_{n-2} + \cdots + c_k a_{n-k}$$

for some constants $c_1, c_2, \ldots, c_k$.

Finding a Close-form Formula for a LHRRWCC

1. Find the characteristic equation for the recurrence relation and find its roots. Use these roots to create the general closed-form solution. If the roots are $r_1, r_2, \ldots, r_k$ the general solution is $a_n = \theta_1 r_1^n + \cdots + \theta_k r_k^n$.

2. Use the base values of the recurrence relation and the general solution to form a system of linear equations. The solutions to this system are the coefficients for the desired specific closed-form solution to the recurrence relation.

Exercises

Solutions can be found at http://www.mathcs.bethel.edu/~gossett/DMGN/.

1. Use the theorem: $\gcd(a, b) = \gcd(a, b \bmod a)$ to write a recursive algorithm to calculate the gcd of nonnegative integers a and b. (See page 52 for details on the mod operator.)

2. The number of ways to choose k items without replacement from a set of n distinct objects is denoted as $\binom{n}{k}$ (see page 105). Assume that both n and k are nonnegative. The following values can be treated as base conditions: $\binom{n}{0} = 1$, $\binom{n}{n} = 1$, and $\binom{n}{k} = 0$ if $k > n$. Pascal's Theorem states that $\binom{n}{k} = \binom{n-1}{k-1} + \binom{n-1}{k}$. If you know pseudocode or a programming language that supports recursion, write a recursive algorithm to calculate $\binom{n}{k}$ for $n, k \geq 0$.

3. Define the function f by

$$f(n) = \begin{cases} \dfrac{n}{2} & \text{if } n \text{ is even} \\ 3n + 1 & \text{if } n \text{ is odd} \end{cases} \quad \text{for integers } n \geq 1$$

Let n_0 be any positive integer and define $n_i = f(n_{i-1})$. The *Collatz conjecture* asserts that $n_i = 1$ for some $i < \infty$.

Write a recursive algorithm that prints the sequence of values generated by f (including the initial n). It should terminate once the value 1 has been printed. (*Print* means the return value will be void.)

4. Find closed-form solutions for the following linear homogeneous recurrence relations with constant coefficients. Use the Rational Roots Theorem (next page) for part (c).

    (a) $a_0 = 1, a_1 = 3, a_n = -a_{n-1} + 2a_{n-2}$ for $n \geq 2$.

    (b) $a_0 = 0, a_1 = 1, a_n = 9a_{n-1} - 20a_{n-2}$ for $n \geq 2$

    (c) $a_0 = 2, a_1 = -1, a_2 = 1, a_n = 3a_{n-1} + 4a_{n-2} - 12a_{n-3}$ for $n \geq 3$

    (d) $a_0 = 3, a_n = 4a_{n-1}$ for $n \geq 1$

    (e) $a_0 = 1, a_1 = 2, a_n = a_{n-2}$ for $n \geq 2$ (pay attention to the missing term)

Characteristic Equations Having Repeated Roots

Earlier in this chapter, Isolde stated that her technique for finding the closed-form formula for a LHRRWCC does not work if the characteristic equation has repeated roots. Here is an example to illustrate where her technique fails.

Consider the following recurrence relation.

$$a_0 = 1$$
$$a_1 = 1$$
$$a_2 = 2$$
$$a_n = 5a_{n-1} - 8a_{n-2} + 4a_{n-3} \text{ for } n \geq 3$$

The characteristic equation is $x^3 - 5x^2 + 8x - 4 = 0$. The following theorem can be used to find the roots for this equation.

Theorem *The Rational Roots Theorem*

Suppose the polynomial $c_n x^n + c_{n-1} x^{n-1} + \cdots + c_1 x + c_0$ has integer coefficients where $c_n \neq 0$ and $c_0 \neq 0$. Then any rational (or integer) zero of the polynomial must be of the form $\pm \frac{p}{q}$, where p evenly divides c_0 and q evenly divides c_n.

For this example, the potential rational roots are: $\pm 1, \pm 2, \pm 4$. These potential roots can be substituted into the equation to see which (if any) are actual roots. In this case, 1 and 2 are roots, with 2 showing up twice.

$$x^3 - 5x^2 + 8x - 4 = (x - 1)(x - 2)^2 = 0$$

Here is the general solution from Isolde's technique.

$$a_n = \theta_1 1^n + \theta_2 2^n + \theta_3 2^n$$

The resulting system of linear equations that use the base conditions is:

$$\theta_1 + \theta_2 + \theta_3 = 1$$
$$\theta_1 + 2\theta_2 + 2\theta_3 = 1$$
$$\theta_1 + 4\theta_2 + 4\theta_3 = 2$$

Subtract the first equation from each of the other two. This produces

$$\theta_2 + \theta_3 = 0$$
$$3\theta_2 + 3\theta_3 = 1$$

Now subtract 3 times the first equation from the second equation to end up with $0 = 1$. So the technique fails miserably. However, writing the general solution in the form $a_n = \theta_1 + (\theta_2 + \theta_3 n)2^n$ will produce a system of linear equations that *does* work.

$$a_n = 2 + \left(-1 + \tfrac{1}{2}n\right) 2^n \text{ for } n \geq 0.$$

Consult your favorite *Discrete Mathematics* textbook for full details.

Chapter 11

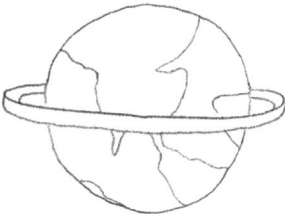

Isolde! I heard that you and Logan went on a date to a concert. I thought you weren't interested in dating.

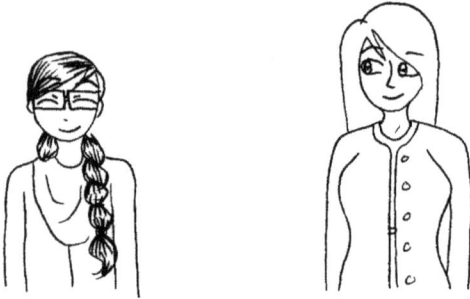

I said it was too soon to think about romance and marriage. I didn't say I plan to become a hermit. Having a social life and doing things with friends is still important to me. Logan and I are good friends, so I enjoy spending time with him.

Ok. Tell me about the concert.

It was a performance by the symphony orchestra and a guest violinist. The program consisted of the two romances for violin and orchestra by Beethoven and then the Dvořák violin concerto in A minor. I really enjoyed the music.

I am so jealous! I love the Beethoven romances! It must be nice to be old enough to go out to concerts and other events whenever you want.

It isn't that simple. University students don't have unlimited amounts of time.

Still, you don't need to get parental approval for everything.

Oh! I have a great math problem that my regular math teacher told us about in class today. He didn't tell us the solution yet, but I am pretty sure I solved it. However, my solution is really surprising, so I thought I would have you look at it and tell me if I made a mistake. Here is the problem:

Suppose you tie a (very long) rope tightly around the equator of the earth. How much longer must the rope become in order to float 1 inch from the surface all the way around the earth?

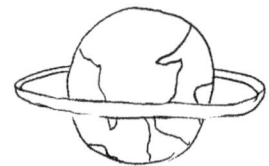

How much rope can I buy for $10.00?

Tell me about your solution.

Well, the first thing I needed to do was find out what the circumference of the earth is. I looked on the web and came up with 24,901.55 miles at the equator. I decided to just think about the equator, so the rope could be imagined to be going around the circumference of a circle.

The circumference of a circle satisfies $c = 2\pi r$, so I found the radius of the earth at the equator to be $r = 3963.204773149987$ miles.

So you are assuming that the earth is a perfect sphere, even though it actually is not. Since you only care about the rope around the equator, I don't see any harm in that assumption.

Here is the idea for which I am proudest. To make the rope float 1 inch off of the surface, I just need to add 1 inch to the radius of the earth at the equator. In order to do that, I needed to convert one inch into miles.

$$1 \text{ inch} \times \frac{1 \text{ foot}}{12 \text{ inches}} \times \frac{1 \text{ mile}}{5280 \text{ feet}}$$

$$= \frac{1}{63360} \text{ mile}$$

$$\simeq 0.000015782828 \text{ mile}$$

So the new radius would be 3963.204788932815 miles. Solving for the new circumference results in

$$c = 2\pi r = 24901.550099166440 \text{ miles}$$

The difference between the new and old circumferences is

$$0.0000991664 \text{ miles} \times \frac{5280 \text{ feet}}{1 \text{ mile}} \times \frac{12 \text{ inches}}{1 \text{ foot}}$$

$$\simeq 6.28 \text{ inches}$$

That is quite amazing, and very counterintuitive! Hmm ... Here is a simple way to validate your result. The new circumference can be found as $c_n = 2\pi(r + 1 \text{ inch})$. Subtracting the old circumference leads to

$$c_n - c_o = 2\pi(r + 1 \text{ inch}) - 2\pi r = 2\pi \text{ inches}$$

Since $2\pi \simeq 6.28$, your answer is correct.

So I did all that work with all those digits after the decimal point for no good reason!

Not really. Sometimes it is best to find a correct solution, then simplify it.

Thinking About Group Projects

During your time as a student, you have almost certainly been asked to participate in a group project. There are many reasons why you have been asked to do so. The following (incomplete) list comes to mind.

- Many work environments are inherently group-based. Working on team projects while in school helps to prepare you for that environment.

- Members of the team can share ideas with each other, potentially producing a better result than if the team members all worked on their own.

- If the project is complex, breaking the class into groups reduces the grading burden on the instructor.

There are also some inherent risks that are associated with group projects. Some of the most common are listed next.

- Some members of the group don't make much of an effort to contribute. The other group members either need to take up the slack, or share in a lower grade.

- Some members of the group do not have course knowledge and sufficient skills to make real contributions. They may have a desire to contribute, but they are not up to the task.

- One of the team members decides that the other members are not able to contribute at a level that would ensure a sufficiently high grade, so that person unilaterally decides to do the entire project. This robs the other team members of the chance to contribute and learn.

Think about the group projects in which you have participated. What worked well? What kinds of problems (with the team dynamics) did you encounter?

Discuss your answers with a few other people. Then, as a group, think about ways to make future projects work better. What can the instructor do to help? What can the team members do to help? Try to make specific suggestions to alleviate the problems mentioned above (as well as other problems you have observed).

Finite-State Machines;
Regular Expressions

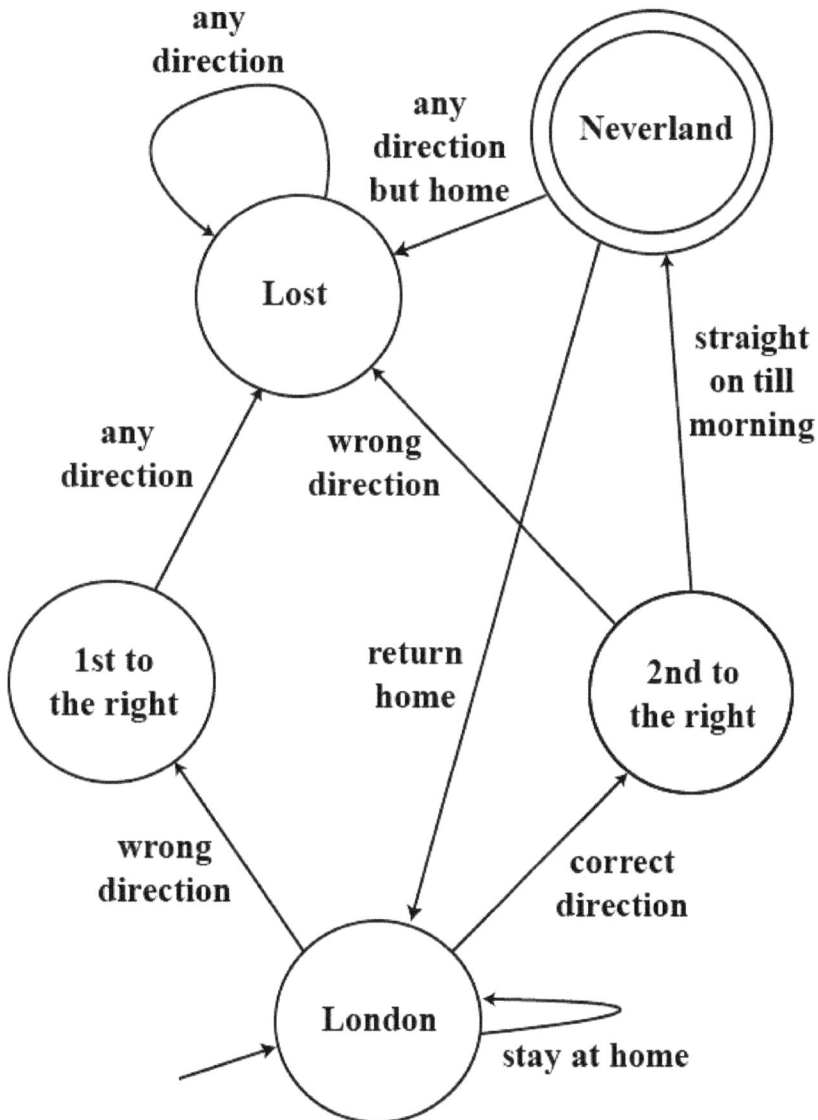

Today's topics come from computer science. The first is a way to model some relatively simple kinds of computation. The model is called a *finite-state machine* or a *finite-state automaton*. One application is creating parts of a compiler (a program that translates human-readable code into machine-readable code). We will look at simpler applications.

This diagram is an example of a finite-state machine. It takes a sequence of digits and determines whether the cumulative sum is even or odd. There are two *states*: "Even" and "Odd", denoted by the circles. The Even state has a double circle, indicating that it is a *final state*.

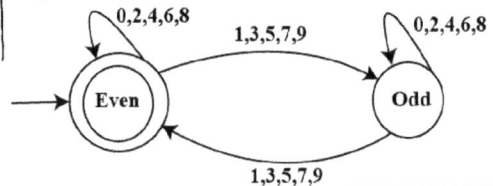

The possible *inputs* are the 10 digits. If the machine is currently in state Even and an odd digit is next in the input sequence, then the machine transitions into the Odd state (as indicated by the arrow labeled with odd digits). On the other hand, if an even digit arrives next, the machine stays in state Even (as indicated by the loop labeled with the even digits). Similar comments apply if the machine is currently in the Odd state. The unlabeled arrow on the left indicates that Even is the *start state*.

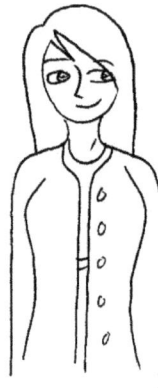

So how does it determine the sum?

It doesn't. Its job is to determine the parity of the current sum. We "win" if it is Even and "lose" if it is Odd.

Here is a formal definition.

Definition *Finite-State Machine*

A *finite-state machine*, A, is a model that consists of

- A finite set of states, $\mathcal{S}$.
- A set of input values, Σ.
- A transition function, $t(s, i)$, that maps state–input pairs to states.
- A special state called the *start state* (generically named s_0).
- A subset, $\mathcal{F} \subseteq \mathcal{S}$, of *final states*.

The definition does not depend on the diagram. In fact, we could specify the machine by using a *transition table*. Also, "final" state does *not* mean "the last state produced by the input sequence".

Here is the same example, specified using a diagram and then using a transition table. The two expressions are equivalent. Humans tend to like the diagram but a transition table is easier to translate into computer code. For the table, we need to explicitly specify the start state and the set of final states.

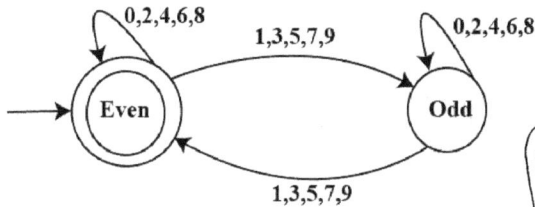

State	Input	
	0,2,4,6,8	1,3,5,7,9
Even	Even	Odd
Odd	Odd	Even

The start state is Even.
The only final state is Even.

Ok, Lily, if you think you understand this, here is one for you. The input set is $\Sigma = \{a, b, c\}$. The start state is s_0, and there are three final states. The first two final states respectively identify an input sequence with a double consonant (bb or cc) as the most recent pair of input symbols. The other final state indicates that the most recent input symbol was a. You may need other states besides the four I identified.

Well, the start state indicates that I have an empty input sequence, so it will not be a final state. If the most recent symbol is a b or c and the previous symbol wasn't the same, then I would not be in a final state, so I need another state to indicate that. Actually, I need two more states: one to indicate that b was just received but not bb, the other for c but not cc.

Here is my diagram.

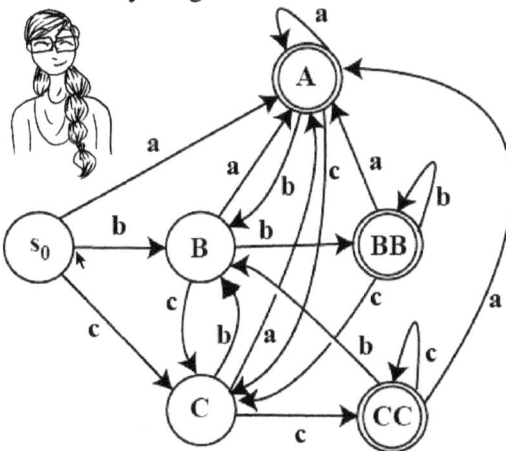

I see two issues:
(1) you forgot the empty arrow entering the start state.

(2) it feels a bit messy. Repositioning a few states may help.

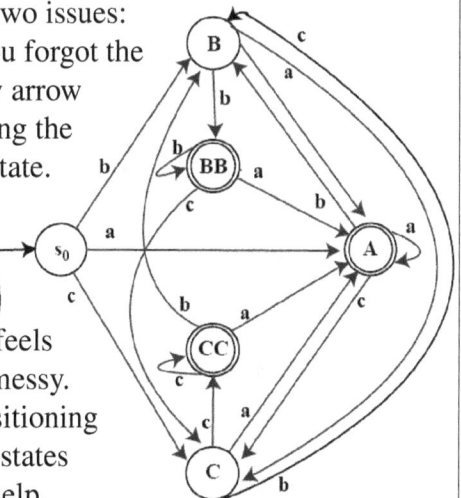

167

The second topic for today is a pattern matching tool called a *regular expression*. Regular expressions provide a mechanism for specifying a group of strings with a desired pattern. A *string* is a finite sequence of characters that are concatenated together.

Suppose, for example, that I want to find all strings in a document that look like a standard license plate in a state where license plates consist of 3 letters, a space, and 3 digits.

ZWE 896
ASD 231

A regular expression to find any such string looks like

```
␣[A-Z][A-Z][A-Z]␣[0-9][0-9][0-9]␣
```

where ␣ represents a space character. [A-Z] represents any uppercase letter between A and Z (inclusive). [0-9] represents any digit between 0 and 9.

To simplify things, I am assuming that license plate strings don't occur at the beginning or end of a line.

How would I specify a string with a square bracket, [, in it? For example:

"I here [sic] that she is ill."

As you have noticed, the square bracket, and several other characters have special roles in regular expressions. These special characters are called metacharacters. They include:

```
$ ^ [ ] | ( ) \ . " * ? +
```

If you wish to specify a metacharacter as part of your target strings, you can do a backslash-escape. For example, \[matches an opening square bracket.

We could specify your example string by using the metacharacter +. That character specifies "one or more" of the character specified just to its left.

```
[A-Za-z ]+\[[a-z]+\][A-Za-z ]+\.
```

The subexpression [A-Za-z] specifies any one character in the set consisting of all the uppercase and lowercase letters and also the space character. The "+" at the end indicates that there may be one or more letters and spaces. The \[subexpression indicates that an opening square bracket must come next, followed by some lowercase letters and then a closing square bracket. Next comes more letters and spaces and then a single period (specified by \.).

So would it be possible to shorten the license plate regular expression to

␣[A-Z]+␣[0-9]+␣

by using the + metacharacter?

It is tempting to try that, but it doesn't work. Although it *will* match any valid license plate string, it also matches strings that we *don't* want. For example,

"A 1352" or "CD 12" or "G 7" or even

"UVWXYZ 1234567890".

The problem is that the + metacharacter can't specify "exactly 3."

It *is* true that most modern computer programming languages support regular expressions. They often add extensions to allow a regular expression to specify something like "exactly 3". However, that is not something we need for this brief overview. I plan to show just a few of the more common ways to specify regular expressions.

The simplest rule in regular expressions is: **most characters match themselves.** The exception to this rule is: **metacharacters don't match themselves.**

So, the regular expression "cat" would match lines containing "concatenate" or "the cat in the hat", etc.

catch a cat
or abdicate

Here are some (but not all) ways to specify regular expressions.

- The $ character matches the end of a line. So, the regular expression cat$ only matches lines where cat are the last 3 characters on the line.

- The ^ character matches the beginning of a line.

- The [character initiates a [] pair. A regular expression consisting of a pair [] with a set of characters inside matches any *one* character from the set.

  A regular expression consisting of [^] with a set of characters after the ^ matches any one character that *doesn't* occur inside the [] pair.

- The metacharacter | indicates an alternative. A regular expression containing a | matches any string that contains either the left or the right alternative.

- The () metacharacters are used to group characters into sub-patterns of the regular expression. For example, the regular expression c(a|u)t matches any string that contains *cat*, or *cut* as a substring. It does not match *caut*.

- The + metacharacter matches *one or more* of whatever subpattern immediately precedes it. The * metacharacter matches *zero or more* of the subpattern that immediately precedes it. The ? metacharacter matches *zero or one* copy of the preceding subpattern.

Here are some examples.

- `^X[a-z]*[0-9]?$` matches: "Xabc4", "X5", "Xmyt", and "X". It does not match: "a3", "Xbc54", "wXn2", nor "Xab4c".

- `b(oo|ee|ea)t` matches "my boot", "I ate beets", and "beat it kid!". It does not match "boo", "bot", nor "bean", nor "beeet".

I have a question. What is the difference between `[ab]` and `(ab)`.

`[ab]` matches either an a or a b, but not both. `(ab)` matches the pair of characters ab.

Can you create an expression to match lines of lowercase letters, semicolons, periods, and spaces?

Maybe `[^0-9]+` because it excludes digits but allows all the characters you specified.

Oh! Wait! It would also allow uppercase letters and semicolons. Too bad.

Perhaps `[a-z ;.]+`

That is close, but it will match the line "4s5" since it matches the "s". You can fix it by using ^ and $ to say that the whole line must *only* include those characters. The correct version is: `^[a-z ;.]+$`

It was good that you included a space character in the list. If you need to have a "−" character inside square brackets, make it the first character. Otherwise it is interpreted as a range operator. If you want ^ as a character in the list, make sure it does *not* come first.

Warning: inside `[]`, metacharacters have no special meaning (except ^ as first character). The pattern `[a\^b]` makes \ one of 4 possible choices; use `[a^b]` to include ^ as one of just 3 choices.

170

Here is another useful rule: the period metacharacter matches any character except a newline. You need to use either `[.]` or `\.` if you want an actual period. As an example,

`.*[A-Z][a-z]* [a-z]+\..*`

matches any line that contains a two-word sentence ending with a period. The `.*` subexpressions at the front and back indicate that anything at all can come before or after that sentence. The regular expression also requires an uppercase letter at the start of the sentence.

-> Match me. Please!

So the regular expression
`.*[A-Z][a-z]* [a-z]+\..*`
would match "I sneezed." and "Hi honey! Hug me." or even "343khf&%My cat.*&r2#1", but not "I'm home." nor "kiss me."

I don't think it is very smart, since it matches "X k."

You are correct. Regular expressions are not smart in the sense that they are only looking for patterns of characters. They have no idea what constitutes a valid English sentence.

That does not mean that they are useless. If you are searching for something in a long text, it is nice to be able to specify a pattern to match instead of an explicit set of characters.

For example, I can use a regular expression to match variations on a title:

`(Dr|Dr\.|Doctor) Who`

allows a single search rather than trying each of the three variants one at a time.

The Tardis

The final thing I want to mention today is a nice theorem. It states that finite-state automata and regular expressions are equally powerful.

Consider a specific finite-state automaton, F. It is always possible (via a constructive algorithm) to create a regular expression that matches all (and only) those input sequences that send F to a final state.

Also, given a specific regular expression, E, it is possible to create a finite-state automaton that ends in a final state for exactly the input lines that E matches.

I feel very expressive. I think I will write a poem.

Definitions

Finite-State Machine A *finite-state machine*, A, is a model that consists of

- A finite set of states, $\mathcal{S}$.
- A set of input values, Σ.
- A transition function, $t(s, i)$, that maps state–input pairs to states.
- A special state called the *start state* (generically named s_0).
- A subset, $\mathcal{F} \subseteq \mathcal{S}$, of *final states*.

String A string is a finite sequence of characters that are concatenated together.

Regular Expression (An Informal Definition of Regular Expressions) Let Σ be an alphabet. A *regular expression* over Σ is a mechanism for recognizing or matching a string from Σ. The subset is called the *regular set* generated by the regular expression. A regular expression serves as an abstract pattern that specifies which strings in Σ^* (the set of all finite-length strings from Σ) belong to the corresponding regular set.

- The metacharacters are: $ ^ [ ] | ( ) \ . " * ? +
- Characters which are not metacharacters match themselves.
- Backslash-escape matches a metacharacter. For example, \\ matches \
- A period matches any character except a newline character.
- The $ character matches the end of a line.
- The ^ character matches the beginning of a line.
- A regular expression consisting of a pair [ ] with a set of characters inside matches any *one* character from the set.
- A regular expression consisting of [^] with a set of characters after the ^ matches any one character that *doesn't* occur inside the [ ] pair.
- The metacharacter | denotes an alternative. A regular expression containing | matches any string that contains either the left or the right alternative.
- The ( ) metacharacters are used to group characters into subpatterns of the regular expression.
- The + metacharacter matches *one or more*. The * metacharacter matches *zero or more*. The ? metacharacter matches *zero or one* copy of the preceding subpattern.

Theorem

Equivalence Finite-State Machines and Regular Expressions are equally powerful. They both can be used to specify any regular set.

Exercises

Solutions can be found at `http://www.mathcs.bethel.edu/~gossett/DMGN/`.

1. Let $\Sigma = \{a, b, c\}$. Create a finite-state machine that matches input sequences that contain a double a anywhere in the sequence (and which does not match any other sequences).

2. Let $\Sigma = \{a, b, c\}$. Create a finite-state machine that matches input sequences that *do not* contain a double a anywhere in the sequence (and which does not match any other sequences).

3. Let $\Sigma = \{0, 1, 2\}$. Create a finite-state machine that determines whether the mod 3 sum of the input sequence is 1. (A mod 3 sum of 1 is the only final state). Example: the mod 3 sum of 2,2,1,0 is 2.

4. Let $\Sigma = \{0, 1, 2\}$. Create a finite-state machine that determines whether the input sequence is divisible by 2. The input sequence is considered to be the digits of an integer, n, arriving in right-to-left order. So the sequence 2,2,1,0 would correspond to $n = 0122$ and should land in a final state since 122 is divisible by 2.

5. Let $\Sigma = \{a, b, c\}$. Which input sequences send the following finite-state machine into a final state?

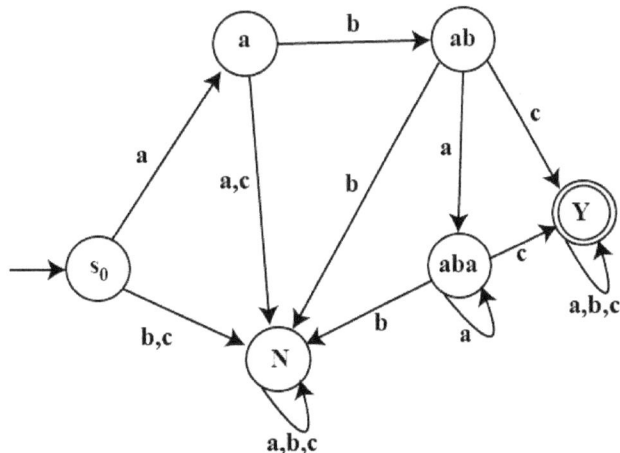

173

You may find it helpful to use the form at
`http://www.mathcs.bethel.edu/~gossett/cgi-bin/test-regex.pl`
to test your solutions to these exercises. Assume that the strings to be matched will not contain any newline characters except possibly at the end of the string.

6. Let $\Sigma = \{a, b, c\}$. Create a regular expression that matches strings that contain a double a (and which does not match any other strings). Use ^ and $.

7. Let $\Sigma = \{a, b, c\}$. Create a regular expression that matches strings that *do not* contain a double a (and which does not match any other strings). Use ^ and $.

8. Let Σ be the full set of characters that are found on a standard keyboard. Create a regular expression that matches (and only matches) strings that contain at least one uppercase letter and at least one digit. Use ^ and $.

9. Let Σ be the full set of characters that can be found on a standard keyboard. Create a regular expression that matches (and only matches) strings that are phone numbers in the form (ddd) ddd-dddd, where d represents a single digit. Use ^ and $.

10. Let Σ be the full set of characters that can be found on a standard keyboard. Create a regular expression that matches (and only matches) strings that contain no uppercase letters nor any odd digits. Use ^ and $.

11. Let Σ be the full set of characters that are found on a standard keyboard. Create a regular expression that matches (and only matches) strings that contain at least one of the letter combinations: ouo, ae, or ei. Use ^ and $.

12. Let Σ be the full set of characters that can be found on a standard keyboard. Create a regular expression that matches (and only matches) strings that contain a lowercase letter followed by three periods and then a space (the string may contain additional characters). Use ^ and $.

Chapter 12

Get a job!

Isolde, I have a question before we begin. Besides becoming a high school teacher, what kinds of careers are available to math majors?

That is a great question. Math is one of the majors that opens lots of doors. The obvious alternative to teaching high school is to become a professor, which would involve doing mathematical research and teaching undergraduate or graduate university students.

What if we drop the uniform convergence assumption?

Let $\epsilon > 0$ be given, then $\exists\, \delta > 0 \ni f$ is

If you don't want to be in academia, there are many other options. One of the best is a career as an actuary. Actuaries work with statistical information to make predictions about future events. For example, insurance companies want to predict future claims by policy holders so that they can set their premium rates high enough to cover their costs, but not so high that other insurance companies steal their customers.

In yearly career ratings, actuary is almost always the top job. It is consistently one of the top 3 jobs.

Actually, one of my mom's friends is an actuary. She says it is a great field for women because career advancement is tied to performance on a series of professional exams.

Another great career for math majors is the field of Operations Research. People working in this field help organizations make optimal use of scarce resources. One of my friends who graduated last year is working on an algorithm to find an optimal location for a medical analysis laboratory that serves multiple hospitals and clinics.

Another application might seek to determine the optimal mixture of ingredients (this week) for an animal feed mixture. The goal would be to minimize the cost while still meeting the nutritional requirements.

Data Analytics is a related job area for math majors.

Get this blood sample to the lab right away!

Operations Research sounds fun. What else is available? Can a major in math be combined with other majors?

There are many ways to combine math with other majors. I have friends who are double-majoring in math and computer science, math and physics or chemistry, math and biology (looking towards a medical degree), and also math and business. Other friends are just doing minors in those other majors. One of my professors said that businesses want to hire people with good problem-solving skills (among other things like good communication skills). Math majors are well-known for being able to solve problems, so they are quite competitive for most jobs in business. Some people go with their hearts and combine math with art or theater or communications.

Math helps you to learn how to think and write clearly. That means it can prepare you to do well in areas that don't involve equations or theorems or proofs. One of the best pastors I have known was a math major as an undergraduate.

The Summer Institute of Linguistics, which trains people to translate the Bible into languages that currently have no written form, likes to hire math majors.

There are other careers that specifically hire people to do various kinds of mathematics. Statisticians work for government agencies, research firms, medical companies and research hospitals, and many businesses. Other mathematicians work with scientists and engineers to solve mathematical problems or create mathematical models.

Wow! That is a lot to think about. Thanks!

Lily, before we begin the lesson, I have a great visual puzzle for you to solve. I found it in an old puzzle book, *Perplexing Puzzles and Tantalizing Teasers*, by our old friend Martin Gardner.

I have used 4 sticks to make a stemware glass that is holding a cherry. Your task is to move just 2 of the sticks so that (1) the glass is upside down and (2) the cherry is outside of the glass.

Solve this before turning the page!

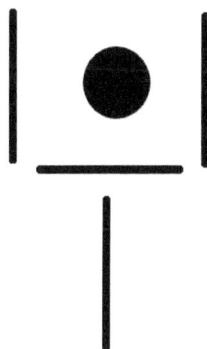

Lily has been thinking for a while.

I know how to turn the glass upside down with only 2 moves. Just move the stem and bottom to the top. But the cherry is still inside.

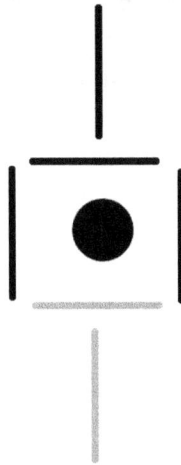

I can't think of a way to get the cherry out of the glass. Do you have any hints?

I'll buy a hint for $50.

We humans like symmetry. Our strong instinct is to place the cherry in an aesthetically pleasing position relative to the upside down glass. Think about other options. Also, do you need to move a stick completely away from its initial position?

Hmm. The initial position hint seems useful. And I will try to not get hung up on symmetry.

Ah! I got it!

I can slide the bottom of the glass a half-stick length to the left. Then move the right edge to become the left edge of the downward facing glass. The old left edge becomes the stem. Turning an edge into the stem was my big breakthrough.

Good work. Let's begin today's lesson now.

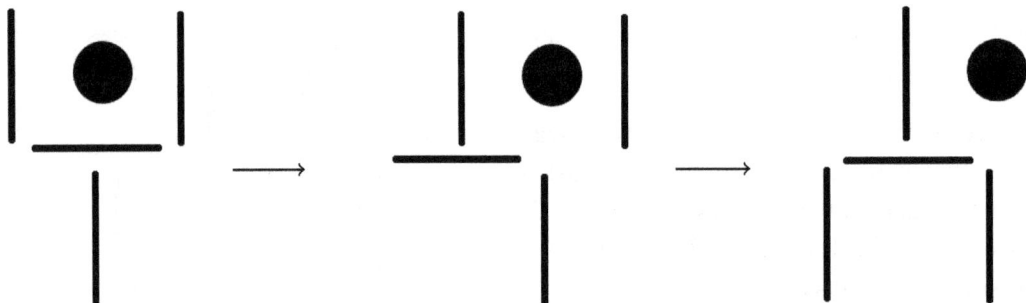

Graphs

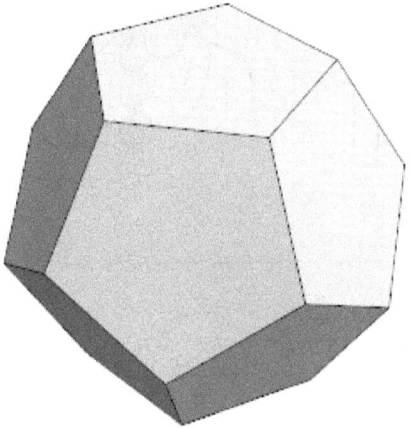

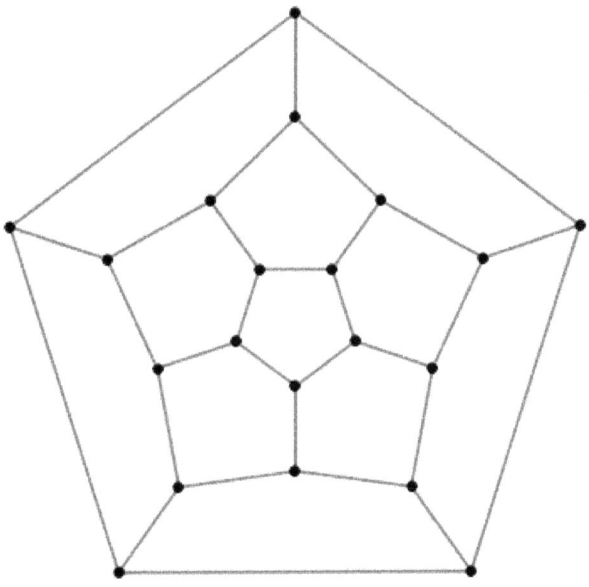

Today I want to introduce you to the topic of graph theory.

Hello graph theory, pleased to meet you.

Actually, I already learned about how to graph functions, and how to make bar charts. What else is there to learn?

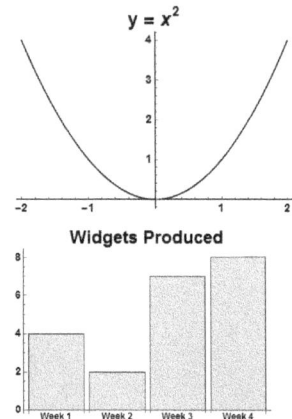

$$y = x^2$$

Widgets Produced

The kind of graph I want to discuss is different. It is a mathematical object that consists of edges and vertices, often represented by lines and points. Edges may intersect at points, and points may be on a common edge. Each edge joins two points. First an example, then a formal definition.

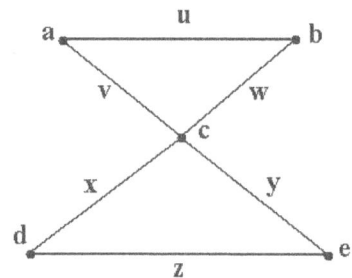

Definition *Simple Graph*

A *simple graph*, $G = (V, E, \phi)$, consists of a nonempty, finite set of *vertices*, V, a finite set of *edges*, E, and a one-to-one incidence function, ϕ, that maps edges to unordered pairs of distinct vertices.

In the previous example, $V = \{a, b, c, d, e\}$, $E = \{u, v, w, x, y, z\}$. One example of the mapping ϕ is $\phi(u) = \{a, b\}$ (this is often written informally as $\phi(u) = ab$).

This definition seems like overkill for defining the picture of a bunch of dots and lines, but there are good reasons for it. The graph is independent of the picture.

The picture is an *embedding* of the graph in the plane. Sometimes the embedding can mislead us. Since the graph is independent of any embedding, the following picture is another representation of the same graph. Notice that the same vertices are at the ends of each edge.

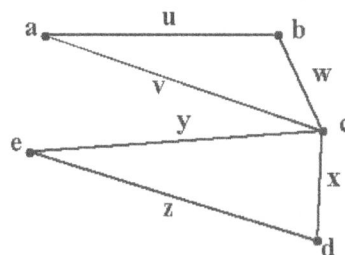

Why attach the word *simple*? Is there another kind called a complicated graph or perhaps a difficult graph?

The alternative is a *graph* (without the adjective), or if we plan to emphasize that it isn't simple, we may call it a *multigraph*.

Definition *Graph*

A *graph*, $G = (V, E, \phi)$, consists of a nonempty, finite set of *vertices*, V, a finite set of *edges*, E, and an incidence function, ϕ, that maps edges to unordered pairs of vertices. If the vertices in the unordered pair are the same vertex, the edge is called a *loop*.

The added features permit multiple edges joining the same pair of vertices (by dropping the *one-to-one* requirement on ϕ) and to have loops (by dropping the *distinct* requirement for the unordered pairs of vertices).

Note: singular – vertex; plural – vertices
vertexes is not a word

So would these be examples of multigraphs?

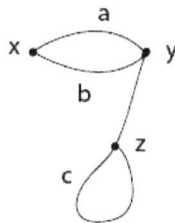

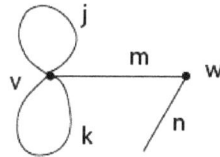

The first one is. The second is not a graph because edge n has only one vertex.

By the way, we often don't bother to label the edges especially in a simple graph (the two incident vertices will uniquely identify the edge).

Here are some additional definitions (I already used one of them).

Definition *Adjacent; Incident*

Two vertices, v and w, are said to be *adjacent* if there is an edge $e \in E$ for which $\phi(e) = vw$. In this case, the vertices v and w are said to be *incident* with e (and vice versa).

Instead of drawing a picture, we can also completely specify a simple graph by its *adjacency matrix*. (Multigraphs can also have adjacency matrices.)

The matrix contains a 1 where two vertices are joined by an edge. Otherwise it contains a 0.

181

Definition *Adjacency Matrix of a Simple Graph*

Let $G = (V, E, \phi)$ be a simple graph with $V = \{v_1, v_2, \ldots, v_n\}$. The *adjacency matrix*, $A = (a_{ij})$, is the n by n matrix whose rows and columns are indexed by the elements of V (in the same order). The element a_{ij} is defined by

$$a_{ij} = \begin{cases} 1 & \text{if } v_i \text{ and } v_j \text{ are adjacent} \\ 0 & \text{otherwise} \end{cases}$$

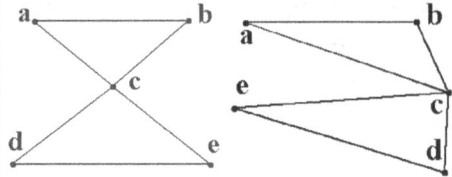

The adjacency matrix for my first example is easy to construct.

	a	b	c	d	e
a	0	1	1	0	0
b	1	0	1	0	0
c	1	1	0	1	1
d	0	0	1	0	1
e	0	0	1	1	0

There are many applications of graphs. One common one is to model computer networks and highway networks.

Graph theory got its start in 1736 when Leonard Euler solved the Königsberg bridge problem. The river Pregel flowed through Königsberg and split into two branches. There were seven bridges over the river. Is it possible to start somewhere in town and walk across every bridge exactly once, ending where you started?

Try to solve this on your own before continuing.

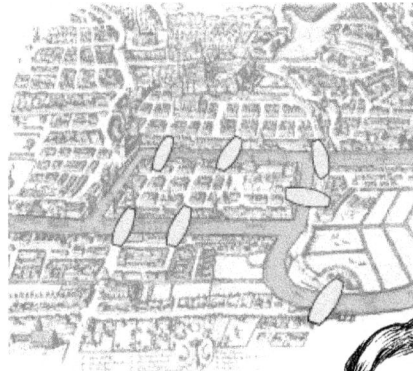

Leonard Euler

Lily, here is a hint. Euler turned the map into a graph - each land mass became a vertex and each bridge became an edge. The result was the Königsberg graph. Notice that it is a multigraph.

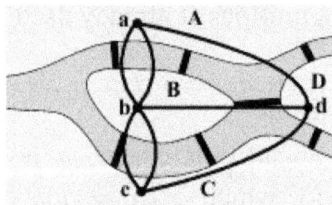

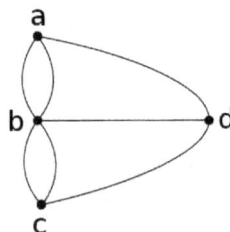

The transition to a graph provided the key insight needed to solve the problem.

182

Ok, that does make things simpler. But I am still stuck.

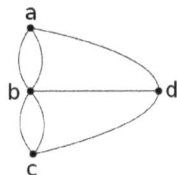

Here is another hint. Imagine starting at a vertex. Every time you leave, you need to come back in order to eventually end at the same vertex. So edges at the initial vertex must come in pairs.

Hmm ... I suppose that implies there can be no edges left over, so the initial node must have an even number of edges.

That is correct. Now think about the other vertices.

For those vertices, we need to come in and then leave (since they are neither the initial nor the final vertex). But every edge needs to be crossed, so none can be left over. Therefore, the vertices come in pairs again. They must all have an even number of edges.

Very good! Notice that your reasoning applies to *any* graph, not just the Königsberg graph. Two useful definitions and a big theorem will complete our visit to the origins of graph theory.

Definition *Degree of a Vertex*

The *degree* of the vertex $v \in V$, denoted $\deg(v)$, is the number of edges that are incident with v. Loops are counted twice (once for each end of the loop).

Definition *Euler Circuit*

Let $G = (V, E, \phi)$ be a graph without loops. An *Euler circuit in G* is a connected path through the graph in which every edge appears exactly one time. The path starts and ends at the same vertex.

Theorem *Euler Circuits*

Let G be a connected multigraph. Then G has an Euler circuit if and only if there are no vertices in G with odd degree.

I didn't discuss the proof of "whenever all vertices have even degree then an Euler Circuit exits". This theorem shows that it is impossible to walk all the bridges of Königsberg exactly once and end up at the same spot.

There is a lot more to graph theory. I will briefly mention two very interesting topics. The first relates to how nice an embedding is possible for a graph and the second is inspired by efficient map coloring.

Consider two possible embeddings for this graph.

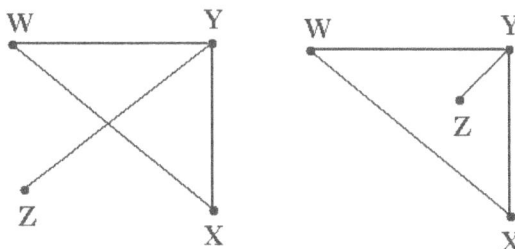

The first embedding has an undesirable (and unnecessary) edge crossing. That is, the edges WX and YZ cross, but as the second embedding shows, that crossing is avoidable.

A graph that can be embedded in the plane without any edge crossing is called a *planar graph*. So, a graph is planar if it has at least one planar embedding.

Here are two graphs that do *not* have any planar embeddings. The first is called *the complete graph on 5 vertices* and is denoted by K_5. The second is *the complete bipartite graph $K_{3,3}$* but is informally called the *utility graph*. (Think about hooking up electricity, gas, and internet service to three houses.)

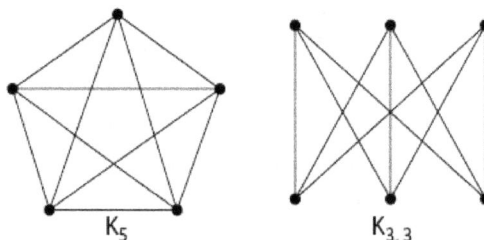

I assume there is a theorem that tells us when a graph is planar and when it is not, but I don't have a clue how such a theorem would be discovered (or stated).

There *is* such a theorem, and it is surprisingly simple. The theorem is Kuratowski's Theorem, and informally, it states that a graph is nonplanar if and only if it contains a subgraph that is a cousin to either K_5 or $K_{3,3}$ (or both). A bit more formally, two graphs are homeomorphic (cousins) if each can be derived from a common ancestor by a (different) sequence of elementary subdivisions. An elementary subdivision looks like adding measles to an edge (the edge splits into two new edges joined by a new vertex).

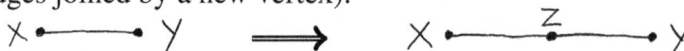

The final topic for today was inspired by the question "what is the minimum number of colors needed to color the countries on a map such that adjacent countries always use different colors".

Consider this map with 4 countries. Using 4 colors certainly guarantees that adjacent countries use different colors. It is easy to see that only 3 colors are required.

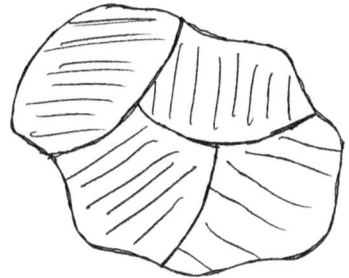

This problem was around for a long time without definitive progress. An early attempt at proving that only 4 colors are ever needed turned out to contain an error. Introducing graphs was one of the first steps to a correct proof (which occurred more than 100 years later and required many hours of computer processing).

Using the Königsberg bridge problem as inspiration, convert each country into a vertex. Make an edge between vertices if their corresponding countries share a common border.

Applying this process to the previous map leads to the following diagram.

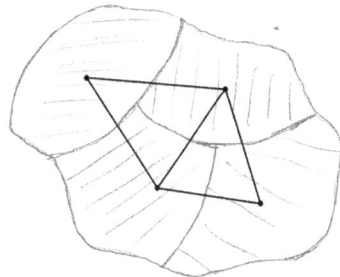

The map coloring can be specified on the graph by attaching colors to the vertices. For this example, we can use Red, Green, and Blue.

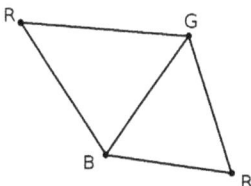

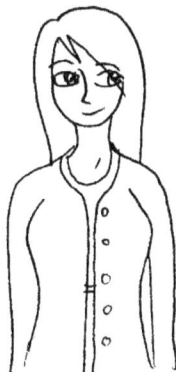

The rule that adjacent countries should have different colors easily translates into a rule that requires adjacent vertices to have different colors. This conversion process will always produce a planar map.

The big questions are these: (1) for a given graph, what is the smallest number of colors required, and (2) is there a maximum number of colors that will suffice for all graphs?

Some definitions will be helpful.

Definition *Chromatic Number*

A *coloring* of a graph $G = (V, E, \phi)$ is an assignment of a color to each vertex in V. The coloring is a *proper coloring* if the two vertices in every pair of distinct adjacent vertices have different colors.

The *chromatic number* of G is the minimum number of colors in any proper coloring. The chromatic number of G is denoted by $\chi(G)$, or just by χ if the choice of G is clear from the context.

Finding χ for a particular graph takes effort. You need to produce an actual coloring using n colors, and then you need to prove that it is impossible to use less than n colors. Once you complete these two steps, you have shown that $\chi = n$.

Lily, How would you prove that $\chi = 3$ for this graph?

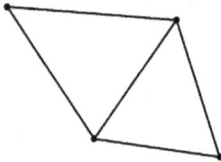

You already produced a coloring using just 3 colors.

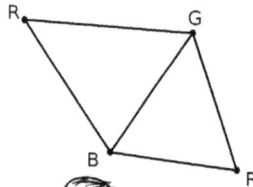

To show that fewer colors won't work, I notice that there is at least one triangle in the graph. To color the vertices on a triangle, I need 3 colors since every vertex is adjacent to the other two.

Since I need at least 3 colors and I have an actual coloring using 3 colors, I have proved that $\chi = 3$ for this graph.

That is correct. Lets end with the famous theorem about chromatic number.

Theorem *The Four-Color Theorem*

The chromatic number of a planar graph is at most 4.

Notice that this theorem only applies to planar graphs. But planar maps convert to planar graphs, so this theorem also works for the original map coloring question. You cannot find a normal map that can't be colored with 4 or fewer colors.

Lily, can you find a graph that requires 5 colors?

Of course, look at K_5. It is nonplanar and needs 5 colors.

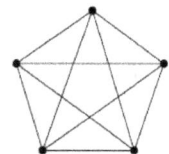

186

Definitions and Theorems

Some of these definitions and theorems are introduced in the exercises.

Definitions

Simple Graph A *simple graph*, $G = (V, E, \phi)$, consists of a nonempty, finite set of *vertices*, V, a finite set of *edges*, E, and a one-to-one incidence function, ϕ, that maps edges to unordered pairs of distinct vertices.

Graph A *graph*, $G = (V, E, \phi)$, consists of a nonempty, finite set of *vertices*, V, a finite set of *edges*, E, and an incidence function, ϕ, that maps edges to unordered pairs of vertices. If the two vertices in the unordered pair are the same vertex, the edge is called a *loop*.

Adjacent; Incident Two vertices, v and w, are said to be *adjacent* if there is an edge $e \in E$ for which $\phi(e) = vw$. In this case, the vertices v and w are said to be *incident* with e (and vice versa).

Adjacency Matrix of a Simple Graph Let $G = (V, E, \phi)$ be a simple graph with $V = \{v_1, v_2, \ldots, v_n\}$. The *adjacency matrix*, $A = (a_{ij})$, is the n by n matrix whose rows and columns are indexed by the elements of V (in the same order). The element a_{ij} is defined by

$$a_{ij} = \begin{cases} 1 & \text{if } v_i \text{ and } v_j \text{ are adjacent} \\ 0 & \text{otherwise} \end{cases}$$

Degree of a Vertex The *degree* of the vertex $v \in V$, denoted $\deg(v)$, is the number of edges that are incident with v. Loops are counted twice (once for each end of the loop).

Bipartite Graph A graph $G = (V, E, \phi)$ is *bipartite* if $E \neq \emptyset$ and if V can be partitioned into two disjoint subsets V_α and V_β such that for every edge $e \in E$, $\phi(e) = v_1 v_2$, where $v_1 \in V_\alpha$ and $v_2 \in V_\beta$.

Euler Circuit Let $G = (V, E, \phi)$ be a graph without loops. An *Euler circuit in G* is a connected path through the graph in which every edge appears exactly one time. The path starts and ends at the same vertex.

Euler Trail Let $G = (V, E, \phi)$ be a graph without loops. An *Euler trail in G* is a path that contains every edge in E exactly once. The initial and final vertices need not be the same.

Planar Graph A graph that can be embedded in the plane without any edge crossing is called a *planar graph*. (A graph is planar if it has at least one planar embedding.)

Chromatic Number A *coloring* of a graph $G = (V, E, \phi)$ is an assignment of a color to each vertex in V. The coloring is a *proper coloring* if the two vertices in every pair of distinct adjacent vertices have different colors.

The *chromatic number* of G is the minimum number of colors in any proper coloring. The chromatic number of G is denoted by $\chi(G)$, or just by χ if the choice of G is clear from the context.

Theorems

Euler Circuits Let G be a connected multigraph. Then G has an Euler circuit if and only if there are no vertices in G with odd degree.

Kuratowski's Theorem A graph, G, is nonplanar if and only if it contains a subgraph that is homeomorphic to either K_5 or $K_{3,3}$.

The Four-Color Theorem The chromatic number of a planar graph is at most 4.

The Handshake Theorem Let $G = (V, E, \phi)$ be a graph with $\epsilon = |E|$ edges. Then

$$\sum_{v \in V} \deg(v) = 2\epsilon$$

Exercises

Solutions can be found at `http://www.mathcs.bethel.edu/~gossett/DMGN/`.

1. Consider the following graph.

   (a) List the degree of each vertex.

   (b) Does the graph contain an Euler Circuit? If so, list one, otherwise indicate why one cannot exist.

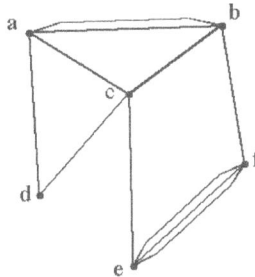

2. Prove the Handshake Theorem.

Theorem *The Handshake Theorem*

Let $G = (V, E, \phi)$ be a graph with $\epsilon = |E|$ edges. Then

$$\sum_{v \in V} \deg(v) = 2\epsilon$$

---

3. Consider the following graph.

   (a) Produce the adjacency matrix.

   (b) Find a planar embedding.

   (c) Find the chromatic number.

   (d) Does the graph contain an Euler Circuit? If so, list one, otherwise indicate why one cannot exist.

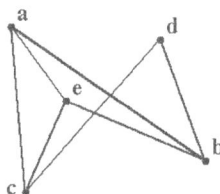

4. Show that the following graph is nonplanar.

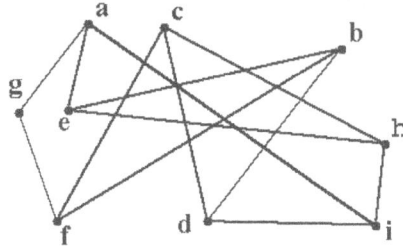

5. Some graphs have the following special property.

> **Definition** *Bipartite Graph*
>
> A graph $G = (V, E, \phi)$ is *bipartite* if $E \neq \emptyset$ and if V can be partitioned into two disjoint subsets V_α and V_β such that for every edge $e \in E$, $\phi(e) = v_1 v_2$, where $v_1 \in V_\alpha$ and $v_2 \in V_\beta$.

Which of the following graphs are bipartite? List the vertex subsets V_α and V_β if the graph *is* bipartite.

G1   **G2**   **G3**

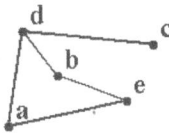 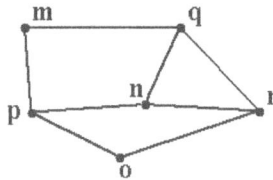 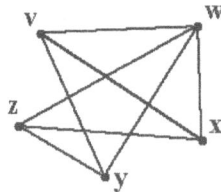

6. Find the chromatic number of each graph in Exercise 5.

7. Find the adjacency matrix for each graph in Exercise 5.

8. An Euler Trail is similar to an Euler Circuit.

> **Definition** *Euler Trail*
>
> Let $G = (V, E, \phi)$ be a graph without loops. An *Euler trail in G* is a path that contains every edge in E exactly once. The initial and final vertices need not be the same.

Create, and prove, a theorem that states when and only when an Euler Trail (but *not* an Euler Circuit) exists in a graph.

Chapter 13

Isolde, I know this is very short notice, but would you be available to come to my piano recital on Friday evening?

Yes, I think I am free that night. What music will you play? Are you the only one performing?

It is a group recital for all my teacher's students. I will perform *Partita No 4*, by Johann Sebastian Bach and a Chinese folk song, 虹彩妹妹 (hóng cǎi mèi mèi), which translates as *Rainbow Girl*, or as *Rainbow Sister*.

Do you play any instruments, Isolde?

No, but I took Irish step dance lessons for about 8 years before I graduated from high school.

My family went to an Irish Fair one year. I enjoyed watching the dancers. Maybe you can dress up and dance for me sometime. (^_^).

I have not danced for a few years. I am sure I am so out of shape that I would not be able to do a decent kick.

Instead of embarrassing myself attempting to step dance, I will offer a puzzle before we start today's session. I found it in an old book my grandmother found at a yard sale. The book is *Number Stories Of Long Ago* by David Eugene Smith and was originally published in 1919.

I will paraphrase the puzzle.

Three men are traveling and meet at dinner time out in the wilderness. They agree to share a meal and want to make equal contributions. One man has 5 loaves of bread, one has 3 loaves of bread, and the other has 8 coins. How should they divide the items in order to do so fairly?

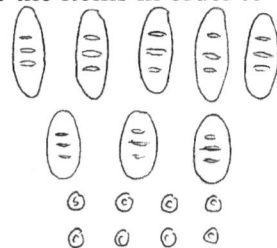

Well, if they divide the 8 loaves evenly, each man will eat $2\frac{2}{3}$ loaves. There are 8 loaves and 8 coins, so each loaf should cost one coin.

Let me name the three men A, B, and C, where A has 5 loaves, B has 3 loaves, and C has 8 coins.

A needs to keep $2\frac{2}{3}$ loaves for himself, so he can sell $2\frac{1}{3}$ loaves to C. B needs to keep $2\frac{2}{3}$ loaves for himself, so he can sell $\frac{1}{3}$ loaf to C. So C should pay $2\frac{1}{3}$ coins to A and $\frac{1}{3}$ coin to B.

But that is a problem, because the coins can't be divided into thirds.

Maybe I am defining "fair" in the wrong way. What assumptions have I made?

Assumption

1 coin = 1 loaf

Suppose I change the assumption that 1 coin equals 1 loaf. What other options are there? ... I think I am stuck.

Part of the 1 coin = 1 loaf assumption is the idea that both the loaves and the coins need to be shared. What if C offers to divide all his coins between A and B in order to get a share of their bread?

It might also be useful to have them divide each loaf into three parts so that we can think about portions instead of fractional loaves.

1 loaf = 3 portions

Ok. If they divide the loaves, there will be 24 portions, so each man should get 8 portions of bread. If C plans to spend all 8 coins, then it makes sense that he pay 1 coin for each of the 8 portions he will receive.

A starts out with 15 portions and can sell 7 portions. B starts with 9 portions and can sell 1 portion. That means A should get 7 coins and B should get 1 coin. Every man gets an equal share of bread and A and B have been paid fairly for the portions they sold.

Correct assumptions are essential for finding a solution.

Here is a related puzzle, from the same source.

A man has 5 short lengths of chain. Each length has 3 links. He wants to join them into a single chain. The blacksmith tells him it will cost $2 to cut a link and $2 to weld a cut link back together. It is much more expensive to create new links for the chain, so the man will have the single chain formed using only existing links, resulting in a chain with 15 links.

What is the least expensive way to accomplish this?

My first thought is to place all 5 small segments in a line. Then you could cut the last link of segment 1 and weld it around the first link of segment 2. Now repeat as you move down the line.

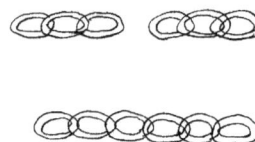

There are 5 small segments, which means there will initially be 4 gaps between segments. To close a gap we need 1 cut and 1 weld, for a total of 4 cuts and 4 welds. At $2 each, that would cost $16.

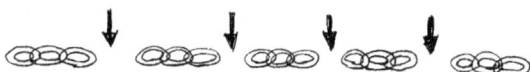

But that seems too straightforward. There must be a better way.

If I don't get the joining-links from the ends of each small segment, they must come from somewhere else. Hmm …

The only other place I can think of would be to break a small segment into 3 single links. But will that be enough links for joining? In the other approach I needed 4 joins.

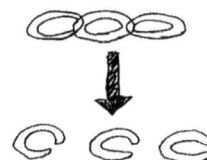

Oh! I see! Once I break a small segment into pieces, there are now only 3 gaps. I can use the 3 open links to do the joins.

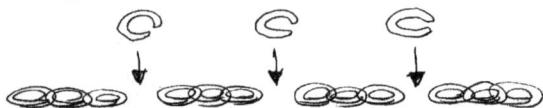

So there are 3 cuts and 3 welds, for a total cost of $12.

Good job! A helpful strategy for finding a best solution is to first find any correct solution, and then look for ways to modify it. You could also look for assumptions that can profitably be changed.

Are you ready to start today's discrete math topic?

Trees

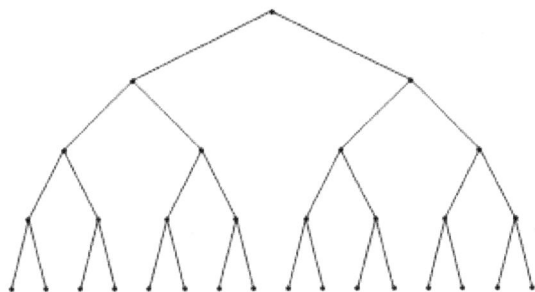

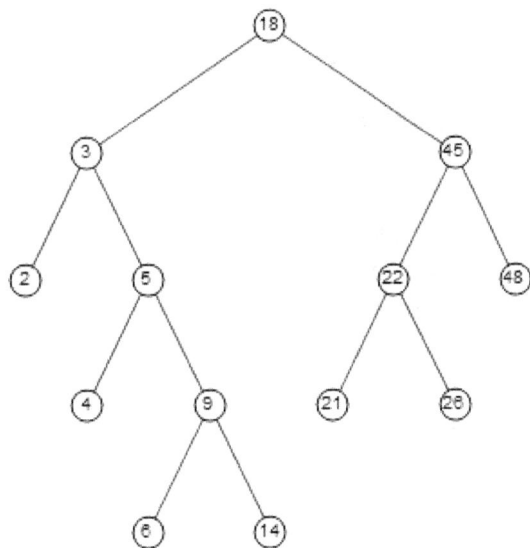

Lily, today I want to do a brief introduction to the topic of trees.

So we are switching from math to forestry?

Not forestry. The topic comes from computer science. Trees are structures that are used to organize data. Trees are a special kind of simple graph. (See page 180 for more about simple graphs.)

Here are the relevant definitions.

Definition *Cycle*

A *cycle* in a simple graph is a non-empty path in the graph that does not contain any repeated edges and which begins and ends at the same vertex.

Definition *Tree*

A *tree* is a connected graph with no cycles.

A graph is *connected* if there is a path between any pair of vertices.

The top graph is a tree, the bottom graph is not.

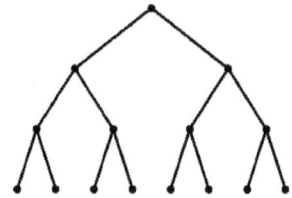

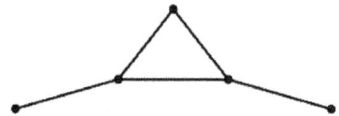

The picture of a tree doesn't look like a real tree.

It actually captures the branching nature of a tree pretty well. It looks wrong because it is usually drawn upside down. In fact, we usually call the vertex at the top the *root* of the tree. If we were to flip it and add a "trunk" edge, it would look more like a real tree.

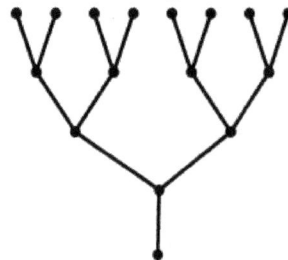

This is *not* how we draw trees.

In the context of trees, the vertices are often called *nodes*. For our sessions, we will always designate one of the nodes to be the root node (this is not always done). In that case, every node, N, (except the root node) has a unique node called its *parent*. The parent of a node, N, is the first node that is encountered on a path from N to the root. The node N is called the *child* of its parent node.

I will restrict our attention to a subset of all possible trees. If each node of a tree can have at most two children, we call it a *binary tree*. In a binary tree, we can designate the children to be *left* and *right* children of their common parent.

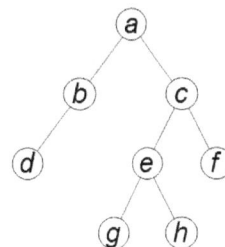

A binary tree.

d is the left child of b
f is the right child of c
e is the parent of g, h
e, g, h form the left subtree of c

In the previous example, d, f, g, and h are called *leaf* nodes because they have no children. All other nodes are *interior nodes*.

Definition *Height; Level*

The *height* of a rooted tree is the length of the longest path from the root to a leaf node.
The *level* of a node in a rooted tree is the length of the unique path from the root to that node. The root node is at level 0.

So the tree in the example has height 3 and node e is at level 2. Other than a lot of new terminology, this seems pretty simple so far.

You are correct. I can stay pretty simple, but perhaps make things a bit more interesting by introducing a couple of counting theorems. Here is the first one.

Theorem *The Number of Edges*

A tree with n nodes has $n - 1$ edges.

Can you prove this Lily? Here is a hint: The theorem indicates that there are almost the same number of nodes and edges. Perhaps you can find a clever way to pair them up (with one leftover node).

Well, I assume that the pairing should not be random. The obvious choice is a paring of nodes with edges to which they are connected.

Can you finish the proof without help from Lily?

There are only two possible pairings of incident nodes and edges: (a) pair a node with an edge leading to a child or (b) pair a node with the edge leading to its parent.

Option (a) seems to be a poor choice for two reasons: (1) If there are two children, it is not clear which child to favor in the pairing, and (2) it is not clear what to do if the node is a leaf node (and there are multiple leaf nodes).

That leaves pairing a node with the edge leading to its parent. Oh! The leftover node would be the root, since it doesn't have a parent!

So here is the proof: Pair each non-root node with the unique edge leading to its parent. That will make $n - 1$ pairs. But every edge connects some node to its parent, so all edges are accounted for. That means there are $n-1$ edges in a tree with n nodes.

You understand that proof completely. I need one more definition before the next theorem.

Definition *Maximal Complete Binary Tree*

A *Maximal Complete Binary Tree of height h* is a binary tree of height h for which every interior node has two children and all leaves appear at level h (the final level).

This tree is a maximal complete binary tree of height 3. It has 15 nodes and 14 edges.

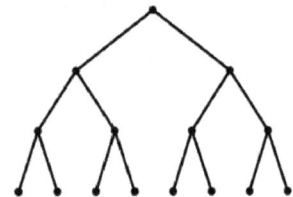

Theorem *Maximal Complete Binary Trees*

Let T be a maximal complete binary tree of height h. Then T has

- $l = 2^h$ leaves

- $n = \frac{2^{h+1}-1}{2-1} = 2^{h+1} - 1$ nodes

- $i = 2^h - 1$ interior nodes

Proof:
At level 0 there is $2^0 = 1$ node. At the next level there will be $2^1 = 2$ nodes, since every interior node has 2 children. At level 2 there will be 2^2 nodes since there are 2 nodes at level 1, each having 2 children. In general, there will be

2^j nodes at level j. There will be $l = 2^h$ leaves, since all leaves in a maximal complete binary tree of height h are at level h.

Thus n will be the sum of a geometric series

$$\sum_{k=0}^{h} 2^k = \frac{2^{h+1} - 1}{2 - 1} = 2^{h+1}-1$$

The number of interior nodes is

$$n-l = \left(2^{h+1}-1\right)-2^h = 2^h-1$$

$\square$

I mentioned earlier that trees are used in computer science to store and organize data. One common way to do that is to store one item at each node. I will first discuss ways to systematically visit every node to access all the data items. Then I will discuss a way to efficiently search the tree for a particular data item.

There are three common methods for traversing all the nodes in a binary tree: preorder, inorder, postorder.

Robot Cats

I preordered a new book I want to read. It will be published next month. (^_^)

Does postorder mean something that comes through the mail?

The terms *pre, in*, and *post* refer to the order in which a parent node is visited, relative to when its children are visited. Consider this tree:

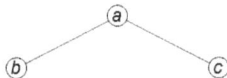

A *preorder* traversal visits the nodes in the order a, b, c (the parent first). An *inorder* traversal visits the nodes in the order b, a, c (the parent in the middle). A *postorder* traversal visits the nodes in the order b, c, a (the parent last). In all cases, the left child is visited before the right child.

Isn't that unfair to the right children?

It *would* be possible to include traversals that visit the right children before the left children, or even to alternate between left and right children in some fashion. However, doing so would not gain any additional applications that are not already possible using the three standard traversals.

These traversals are pretty easy to do. The only complication comes in when the tree has more than just a single parent node. In that case, we can use a recursive algorithm (see page 150). When it is time to visit a node, we consider the left and right subtrees, not just the child nodes. An example should make this clear.

preorder parent, left subtree, right subtree

inorder left subtree, parent, right subtree

postorder left subtree, right subtree, parent

Let's go back to a previous example. A preorder traversal will visit a first, then process the left subtree. The parent of the left subtree is b, which is visited next. The left subtree of b has only d which is treated as a parent node. The right subtree of a is visited next, starting with c, the parent node.

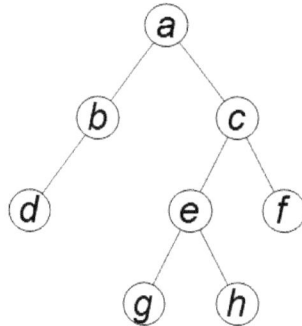

The left subtree of c is next, starting with the parent, e. The left subtree of e has g as parent, so g is next. Then the right subtree of e will mean h is visited next. Finally, the right subtree of c causes f to be visited. So the traversal order is:
$$a, b, d, c, e, g, h, f.$$

I think I can do the inorder and postorder traversals. The inorder will visit the left subtree first so it looks at b. But b is the parent of the left subtree, so it must wait for *its* left subtree to be visited. That means that d actually is the first node to be processed. Then b and then a.

The right subtree if a is now up for processing. We start with the subtree at e but immediately move to g. Then e and then h. The left subtree of c is now done, so c is visited and finally f. The resulting order is: d, b, a, g, e, h, c, f.

The postorder traversal will leave the parent for last, so after visiting d and b as in the inorder traversal, the root node, a, must wait for the right subtree to be visited. The left subtree of c has e as its parent so the subtree below e comes first, resulting in g being the next node to process. Then we skip e for the moment and visit h. Now e gets processed. The right subtree of c is next, resulting in f being visited. Then c and finally a. The traversal is thus: d, b, g, h, e, f, c, a.

preorder : abdceghf

inorder : dbagehcf

postorder : dbghefca

The final topic for today is *binary search trees*. Binary search trees organize data by using a *search key*. The key is something like a number or word that can be sorted, relative to other keys.

Information is added to the tree in a manner that allows subsequent efficient retrieval by using the sort characteristics of the key. This tree is an example. The keys are names of fruits, and are sorted alphabetically.

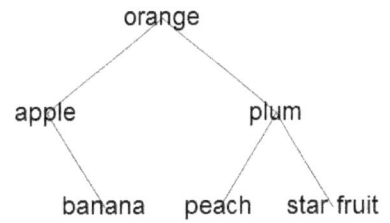

orange

apple   plum

banana   peach   star fruit

This is making me hungry!

The tree is built one node at a time. The fruit tree was built by adding nodes in the key order: orange, apple, banana, plum, star fruit, peach. The first key becomes the root node. Successive keys travel down the tree until they find an open spot for them to be placed. The path traveled is determined by the sort order. If the new key is smaller than the current node, they move towards the left child. Otherwise they travel towards the right child. (We will assume there are no new keys that equal a previous key.) Once an empty spot is found, a new node is added.

We could get the same tree by adding in the order: orange, plum, apple, peach, banana, star fruit.

Once the tree is built, it is easy to determine whether a particular key is present. We follow the same rules to traverse down the tree. We keep moving until we either find the desired key or we hit a leaf node that does not match (in which case we conclude that the key is not in the tree).

Lily, what would be the paths when searching for "banana" and "papaya"?

I would first compare *banana* to *orange* and go to the left child since *banana* comes before *orange*. *Banana* comes after *apple* so I would go to the right child and find the key I want.

With *papaya* I would go right from *orange*, left from *plum*, and finally left from *peach* where I find an empty node. So *papaya* is not in the tree.

201

Finding keys in a binary search tree is quite efficient. The longest search path will be equal to the height of the tree. In the previous fruit tree example, at most three key comparisons are required.

However, if the tree was built in the order: apple, banana, orange, peach, plum, star fruit, it is possible that 6 comparisons would be required.

apple
banana
orange
peach
plum
star fruit

So the order in which we build the tree matters. It looks like we want to avoid alphabetical order.

Yes, alphabetical order (or reverse alphabetical order) results in an inefficient tree. Fortunately, computer scientists have devised tree-building algorithms that produce search trees that are more nearly balanced, so the search paths are close to minimal length.

Recall that a maximal complete binary tree is as balanced as possible. It has minimal height for the number of nodes in the tree.

A theorem (page 198) we saw earlier says that in such a tree, $n = 2^{h+1} - 1$. Solving for h leads to $h = \log_2(n+1) - 1$. Even if some of the nodes on the bottom level are missing, this formula needs only minor revision. I can formalize this.

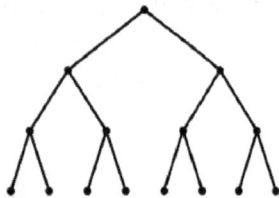

Yes! Please formalize it!

Definition *Complete Binary Tree*

A binary tree of height h is a *complete binary tree* if all leaves appear at levels $h-1$ and h and each interior node has two children.

Theorem *The Height Of A Complete Binary Tree*

If T is a complete binary tree with n nodes, then its height, h, satisfies
$$h = \lceil \log_2(n+1) - 1 \rceil$$

One application of this theorem is that searching for a key in a complete binary search tree is in $\theta(\log_2(n))$ (see page 143).

Ok, Lily, here is a complete binary tree. It has 12 nodes. What does the theorem say about the height of the tree?

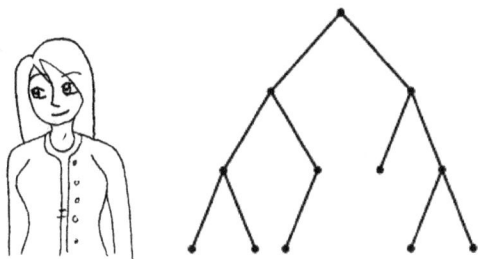

Let me grab my calculator. I think I will need to use the change of base theorem (see page 229). The answer should be 3.

$$h = \lceil \log_2(12+1) - 1 \rceil$$
$$= \left\lceil \frac{\ln(13)}{\ln(2)} - 1 \right\rceil$$
$$= \left\lceil \frac{2.564949357}{.69314718} - 1 \right\rceil$$
$$= \lceil 3.700439718 - 1 \rceil$$
$$= 3$$

Your calculations are correct. The point is that a complete binary search tree has very efficient search time because the height of the tree is kept at a minimum. (I should mention that there are other, less strict definitions of *balanced*.)

Here is one last application of binary search trees. Start with a list of keys in random order and build a binary search tree. Then do an inorder traversal of the tree. The result is interesting.

Try this with the list: f, b, w, c, t, a, k.

Ok, the first step is to build the search tree.

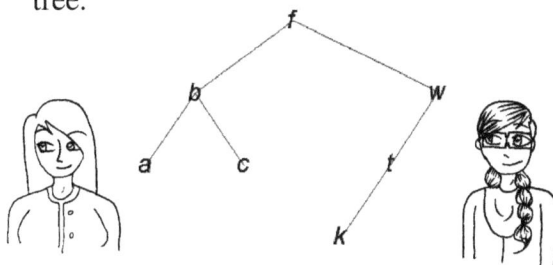

The inorder traversal produces the list:

a b c f k t w.

I see what happened—the list changed from random order to sorted order.

Yes. Binary search trees can be used for sorting. Building a binary search tree with n nodes is typically in $\Theta(n \log_2(n))$ and the traversal is then typically in $\Theta(\log_2(n))$. This algorithm for sorting is therefore in $\Theta(n \log_2(n))$ (the slower phase dominates the performance).

There are many more applications of trees, but I think that is enough for today.

Ok. I can talk with Logan if I want to learn more applications of trees. I am sure he must use them all the time.

Definitions

Cycle A *cycle* in a simple graph is a non-empty path in the graph that does not contain any repeated edges and which begins and ends at the same vertex.

Tree, Node A connected graph with no cycles is called a *tree*. In the context of trees, the vertices are often called *nodes*.

Root, Parent, Child We will always designate one of the nodes to be the root node. In that case, every node, N, (except the root node) has a unique node called its *parent*. The parent of a node, N, is the first node that is encountered on a path from N to the root. The node N is called the *child* of its parent node.

Binary Tree If each node of a tree can have at most two children, we call it a *binary tree*. In a binary tree, we can designate the children to be *left* and *right* children of their common parent.

Maximal Complete Binary Tree A *Maximal Complete Binary Tree of height h* is a binary tree of height h for which every interior node has two children and all leaves appear at level h (the final level).

Binary Search Tree A *binary search tree* is a binary tree used to organize and store data that can be lexicographically ordered. It is defined by the property that the data in a node's left child must precede the node's data in lexicographical order and the data in the right child must lexicographically follow the node's data.

Complete Binary Tree A binary tree of height h is a *complete binary tree* if all leaves appear at levels $h - 1$ and h and each interior node has two children.

Theorems and Traversals

The Number of Edges A tree with n nodes has $n - 1$ edges.

Maximal Complete Binary Trees A maximal complete binary tree of height h has:
$l = 2^h$ leaves, $\qquad n = 2^{h+1} - 1$ nodes, and $\qquad i = 2^h - 1$ interior nodes.

The Height Of A Complete Binary Tree If T is a complete binary tree with n nodes, then its height, h, satisfies
$$h = \lceil \log_2(n + 1) - 1 \rceil$$

Traversals
> **preorder** parent, left subtree, right subtree
> **inorder** left subtree, parent, right subtree
> **postorder** left subtree, right subtree, parent

Exercises

Solutions can be found at `http://www.mathcs.bethel.edu/~gossett/DMGN/`.

1. Is the following graph a tree? Explain your answer.

   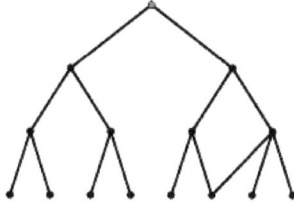

2. (a) How many nodes, leaves, and interior nodes does a maximal complete binary tree of height 5 contain?

   (b) What is the height of a maximal complete binary tree having 4095 nodes? How many edges does this tree have?

3. List the order in which the nodes are visited in this tree when using:

   (a) preoder traversal

   (b) inorder traversal

   (c) postorder traversal

   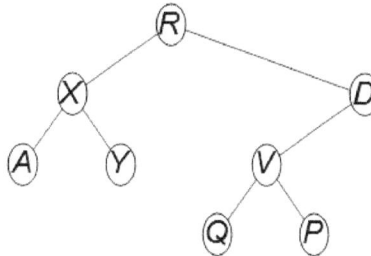

4. List the order in which the nodes are visited in this tree when using:

   (a) preoder traversal

   (b) inorder traversal

   (c) postorder traversal

   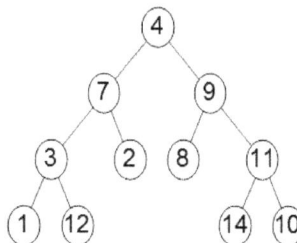

5. For each of the following key sequences, build a binary search tree.

   (a) v, g, h, e, p, w, b, q

   (b) cat, fish, dog, mouse, rat, hamster, lizard, puma

   (c) N, V, S, Y, P, K, Z, A, C, B

6. Suppose T is a complete binary search tree having 1 million nodes. What is the maximum number of nodes that need to be visited in order to find a given key (or determine that the key is not present)?

7. Is this tree a binary search tree?

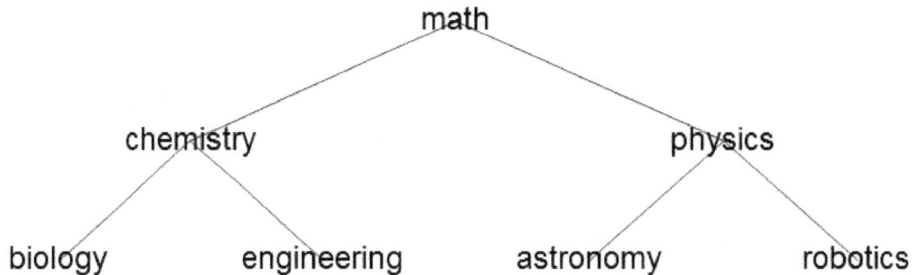

8. A *full binary tree* is a binary tree for which every interior node has exactly two children. (The tree in exercise 7 is a full binary tree.) Prove the following theorem.

   > **Theorem** *Nodes In Full Binary Trees*
   > A full binary tree with n nodes, has
   >
   > - $i = \frac{n-1}{2}$ interior nodes and
   >
   > - $l = \frac{n+1}{2}$ leaves.

9. Prove that if you add an edge to a tree but do not add any new nodes in the process, then the result will no longer be a tree.

Chapter 14

you kids divrde up the christmas gifts on your own.

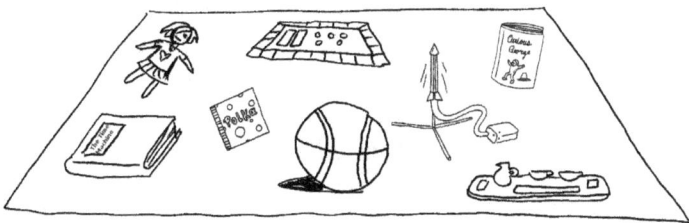

Hi Lily. Would you like to briefly consider another topic before we begin the real lesson for today? It would be about fair division.

Fair division? Would an example of unfair division be when you cheat and make the remainder larger than the divisor? Something like $14 = 3 \cdot 3 + 5$?

No, that would just be an error. (^_^) Fair division is about giving each of several people a fair share of some set of items.

I thought about this after reading a famous story about a wise man.

The wise man was named Solomon. He was the king of Israel around 950 BCE. One day Solomon was hearing grievances from his subjects. Two women came forward with a dispute. They shared a living area and each had given birth at about the same time. But now one of the infants was dead and both were claiming to be the mother of the living child. Solomon's solution was to have a sword brought. He said he would cut the baby in half and give each woman one half. One of the women agreed to this solution but the other pleaded for the baby to be given to her rival. Solomon then knew which woman was the real mother.

Bring me a sword.

The problem we will examine is not as dramatic, but it involves a clever approach to a solution. The division problem concerns a collection of Christmas presents and three siblings. One of their grandmothers is getting on in years. She purchased a number of presents, but didn't indicate who was to receive each present.

The allocation of presents is left up to the children. How can this be done in a manner which the children all feel is fair?

Won't the answer depend on what the presents are and also on the preferences of the kids?

The solution algorithm (called the *method of markers*) is driven by the preferences of the people who are dividing the items.

Let's assume that the children are 13-year-old Arnie, 5-year-old Betty, and 8-year-old Colin.

Grandma has bought the following presents: a doll, a basketball, two books (*Curious George* and *The Time Machine*), a CD (*Polka Favorites*), the game of *Monopoly*, a toy tea set, and a *Stomp Rocket*.

The first step is to line up the objects in some order. Suppose grandma lines them up in the order shown.

doll
basketball
Curious George
Polka Favorites
The Time Machine
Monopoly
tea set
Stomp Rocket

The three children each divide the list into three groups of presents. They do this so that each group has approximately equal value as a potential collection of presents. They can draw two lines to separate the group-ings. The table shows the choices the children have made.

Arnie	Betty	Colin
doll	doll	doll
basketball	————————	basketball
————————	basketball	*Curious George*
Curious George	*Curious George*	————————
Polka Favorites	*Polka Favorites*	*Polka Favorites*
The Time Machine	*The Time Machine*	*The Time Machine*
————————	*Monopoly*	*Monopoly*
Monopoly	————————	tea set
tea set	tea set	————————
Stomp Rocket	*Stomp Rocket*	*Stomp Rocket*

Arnie has no interest in the doll, but would like the basketball. He is a science fiction fan, so *The Time Machine* is attractive. He puts *Curious George* and the CD with the book because he values the basketball higher than the science fiction book. Monopoly is the only present that is of interest in the third group, but it has nearly equal weight with the other groups.

Betty is interested in the doll and the tea set, so they are pretty much alone in groupings. Only *Curious George* is of high interest in the middle grouping, but *Monopoly* is somewhat interesting.

Colin really likes *Curious George* and the *Stomp Rocket*. The middle grouping consists of items that are together so they don't lower his chances of the attractive presents.

The algorithm finds the marker that appears first in the list. In this example, it will be Betty's line below the doll. This indicates that Betty feels the set of presents above the line constitute a fair collection. So Betty gets the doll. Her column is removed, or we may just remove her other marker(s).

The next step sets the pattern for the remaining people.

We locate, among the remaining markers, the one that appears first in the list. In this case it is Arnie's line below the basketball. Arnie will get his second group of presents (those *between* his first and second marker.

Notice that this leaves the basketball unattached. Ignore it for the moment. Now remove Arnie's column. Remove Colin's line that appears inside Arnie's group of presents.

Betty: Doll   Unclaimed: basketball

Arnie: *Curious George, Polka Favorites, The Time Machine*

We now look for the first remaining marker. It is Colin's line that appears below the tea set. Colin will get his third group of presents (the *Stomp Rocket*). This is a collection that he thinks is fair.

There are unclaimed gifts again: the *Monopoly* game and the tea set.

There are multiple ways that the three unclaimed gifts can be handled.

I think I have a good idea about how to distribute the remaining presents in this example. Betty is the only one interested in the tea set. Arnie probably cares more about the basketball than *Monopoly*. Colin will probably be happy with the game (but might fuss a bit about wanting the basketball).

If we use your suggestion, Lily, the final allocation of presents is then

Arnie	Betty	Colin
Curious George	doll	*Stomp Rocket*
Polka Favorites	tea set	*Monopoly*
The Time Machine		
basketball		

Every child has received what they consider to be a fair group of presents and each has actually received a desirable bonus present.

This doesn't seem like a perfect algorithm, but it does seem pretty effective. I suspect that when personal preferences are involved, perfection is too much to hope for.

Functions and Relations

$$\mathcal{D} = \{\diamondsuit, \spadesuit, \heartsuit\}$$

$$\mathcal{R} = \{\infty, \aleph, c\}$$

$$\mathcal{D} \times \mathcal{R} = \{(\diamondsuit, \infty), (\diamondsuit, \aleph), (\diamondsuit, c)$$
$$(\spadesuit, \infty), (\spadesuit, \aleph), (\spadesuit, c)$$
$$(\heartsuit, \infty), (\heartsuit, \aleph), (\heartsuit, c)\}$$

A function from $\mathcal{D}$ to $\mathcal{R}$

$$\{(\diamondsuit, \aleph), (\spadesuit, \aleph), (\heartsuit, \infty)\}$$

A relation between $\mathcal{D}$ and $\mathcal{R}$

$$\{(\diamondsuit, \aleph), (\spadesuit, \aleph), (\spadesuit, c), (\heartsuit, \infty), (\heartsuit, c)\}$$

Today's topic will be functions and relations. And before you ask, *relations* does not mean your aunt Sue and your cousin Tobias.

I already know about functions but haven't learned about relations.

There may be some material about functions that you have not seen yet.

The following definition is necessary in order to present the definition of a function.

This definition also applies when either A or B is an infinite set.

Definition *Cartesian Product*

Let A and B be sets. The *Cartesian product $A \times B$* is the set of all ordered pairs, (a, b), with first coordinate a a member of A and second coordinate b a member of B. That is,

$$A \times B = \{(a, b) \mid a \in A, b \in B\}$$

For example, let $A = \{v, w\}$ and $B = \{1, 2, 3\}$. Then
$A \times B = \{(v, 1), (v, 2), (v, 3), (w, 1), (w, 2), (w, 3)\}$
$B \times A = \{(1, v), (1, w), (2, v), (2, w), (3, v), (3, w)\}$.

Here are the primary definitions.

Definition *Function; Domain; Codomain; Range*

A *function* from the nonempty set $\mathcal{D}$ into the nonempty set $\mathcal{R}$ is a subset, $\mathcal{F}$, of the Cartesian product $\mathcal{D} \times \mathcal{R}$ such that every element of $\mathcal{D}$ appears as the first coordinate in one and only one ordered pair in $\mathcal{F}$.

The set, $\mathcal{D}$, is called the *domain* and the set, $\mathcal{R}$, is called the *codomain*. The *range* of the function is the subset of $\mathcal{R}$ consisting of elements that actually appear in the right-hand side of at least one ordered pair in $\mathcal{F}$.

In the previous example, we could take $\mathcal{D}$ to be A and $\mathcal{R}$ to be B. One possible function would be $\mathcal{F} = \{(v, 2), (w, 1)\}$. However, $\mathcal{G} = \{(v, 2), (v, 1), (w, 1)\}$ would *not* be a function because it violates the "every element of $\mathcal{D}$ appears as the first coordinate in one and only one ordered pair in $\mathcal{G}$" rule.

Hold on! I learned that a function is a mapping from $\mathcal{D}$ to $\mathcal{R}$. What's with all this Cartesian product stuff?

x	$f(x)$
v	2
w	1

The definition you learned was probably something like: *a function is a mapping from the domain to the range such that no element of the domain gets mapped to more than one element of the range.*

To see that the two definitions are equivalent, notice that the phrases "every element of $\mathcal{D}$ appears as the first coordinate in one and only one ordered pair in $\mathcal{F}$" and "no element of the domain gets mapped to more than one element of the range" are identical restrictions on the association between elements in the domain and range. My definition specifies that association by using ordered pairs. You are probably thinking about a table with x and $y = f(x)$ values. Both accomplish the same task.

Ok. I see that they are the same, but I think what I learned is much simpler. Why try to make things more complicated?

The payoff for the added complexity will become clear when I define relations. For now, let's just tolerate the new version of the definition. I have more to say about functions.

There are two optional, deluxe features that a function might exhibit. These special functions are called *one-to-one* functions and *onto* functions. It is also possible for a function to be both one-to-one and onto.

$1-1$

$onto$

Definition *Onto and One-to-One Functions*

Let $\mathcal{F}$ be a function from $\mathcal{D}$ into $\mathcal{R}$. Then $\mathcal{F}$ is called *onto* if every element of $\mathcal{R}$ appears as a second coordinate in at least one ordered pair in $\mathcal{F}$. If no element of $\mathcal{R}$ appears as a second coordinate in more than one ordered pair in $\mathcal{F}$, then $\mathcal{F}$ is called *one-to-one*. (One-to-one is often abbreviated as 1-1).

$\mathcal{F}$ is a 1-1 function. The function $\mathcal{H}$, defined as $\mathcal{H} = \{(v, 2), (w, 2)\}$, is *not* 1-1 because 2 shows up more than once as a second coordinate. It *is* a function, because there are no repeated first coordinates.

The function $\mathcal{F} = \{(v, 2), (w, 1)\}$, mentioned previously is *not* an onto function because no ordered pair includes 3 as a second coordinate.

This is all sort of making sense, but I am still more comfortable with writing $f(v) = 2$ rather than something like $(v, 2) \in \mathcal{F}$. Is that a problem?

Not at all. Since $(v, 2)$ is a point on the graph of f if $f(v) = 2$, they are equivalent approaches. There are two more ideas about functions that we need to consider. The first is the notion of an inverse function. This only works for 1-1, onto functions.

Definition *Inverse Function*

Let $\mathcal{F}$ be a one-to-one and onto function with domain $\mathcal{D}_\mathcal{F}$ and range $\mathcal{R}_\mathcal{F}$. A function, $\mathcal{G}$, whose domain is $\mathcal{R}_\mathcal{F}$ and whose range is $\mathcal{D}_\mathcal{F}$ is called the *inverse of $\mathcal{F}$* if the following conditions hold:

- If $(x, y) \in \mathcal{F}$, then $(y, x) \in \mathcal{G}$.
- If $(y, x) \in \mathcal{G}$, then $(x, y) \in \mathcal{F}$.

Let's express this with your more familiar notation. Functions f and g are inverses if the following conditions are met:

- The domain, $\mathcal{D}_f$, of f is the same as the range, $\mathcal{R}_g$, of g.
- The domain, $\mathcal{D}_g$, of g is the same as the range, $\mathcal{R}_f$, of f.
- For all x in $\mathcal{D}_f$, $g(f(x)) = x$.
- For all y in $\mathcal{D}_g$, $f(g(y)) = y$.

The example function f we have been looking at is 1-1 and onto if we change its codomain to $C = \{1, 2\}$. It must have an inverse, g. The two functions have the following definitions. But why insist on 1-1 and onto?

x	$f(x)$		y	$g(y)$
v	2		2	v
w	1		1	w

Consider the function h that maps $\{1, 2, 3\}$ onto $\{a, b\}$ defined by this table.

x	$h(x)$
1	a
2	b
3	a

If the hoped for inverse function is named k, then we have both $k(a) = 1$ and $k(a) = 3$. But this means that k is not a function, because it maps an element from its domain to two different elements of its range. I will let you think about why onto is required.

Got it!

You have probably already encountered the next definition. I will present the formal definition, then you can translate into the informal notation.

Definition *Composition of Functions*

Let $\mathcal{F}$ be a function whose domain is $\mathcal{X}$ and whose range is $\mathcal{Y}$. Let $\mathcal{G}$ be a function whose domain is $\mathcal{Y}$ and whose range is $\mathcal{Z}$. The composition of $\mathcal{G}$ and $\mathcal{F}$ is denoted by $\mathcal{G} \circ \mathcal{F}$ and is defined by
$$\mathcal{G} \circ \mathcal{F} = \big\{(x, z) \mid \exists y \in \mathcal{Y}$$
$$\text{with } (x, y) \in \mathcal{F} \text{ and } (y, z) \in \mathcal{G}\big\}$$

I remember this from my algebra class. Here is my informal definition:

Let f be a function whose domain is X and whose range is Y. Let g be a function whose domain is Y and whose range is Z. The composition of g and f is denoted by $g \circ f$ and is defined by $(g \circ f)(x) = g(f(x))$. Here's an example.

Let $X = Y = Z = \mathbb{R}$. Set $f(x) = 2x + 1$ and $g(y) = 3y + 5$ and let $h = g \circ f$. Then $z = h(x) = (g \circ f)(x) = g(f(x)) = 3(2x + 1) + 5 = 6x + 8$.

You have a good memory! Here is how I like to interpret composition of functions. Suppose you live in country X, in city x. You would like to travel to city z in country Z. One route is to take the f train from x into the city y in country Y, then transfer to the g train which takes you to z. The alternative route is to take the $g \circ f$ airplane that flies directly from x to z.

It is time to move on to relations.

Relations are generalizations of functions. That means that every function will also be a relation, but some relations will not be functions. Relations are what we get if we eliminate the requirement that every element of the domain can only be paired with one element of the range.

Here is an example of a relation. Let the domain be all the students in one of your classes. We want to match each of them with their siblings. You and I would have

no problem—I have no siblings and you have only one. But some students will have several. Including both (Betty, Fritz) and (Betty, Petunia) would be illegal in a function, but acceptable in a relation.

So this example is a relation formed from our family relations. (^_^)

Here is the formal definition.

Definition *Relation*

A *relation* between the set $\mathcal{A}$ and the set $\mathcal{B}$ is a subset, $\mathcal{R}$, of the Cartesian product $\mathcal{A} \times \mathcal{B}$.

If $(a, b) \in \mathcal{R}$, it is common to write $a \mathcal{R} b$ and to say that a is related to b. If $\mathcal{A} = \mathcal{B}$, the relation is said to be a *relation on* $\mathcal{A}$.

The set $\mathcal{A}$ is called the *domain* of the relation and the set $\mathcal{B}$ is called the *codomain*.

If we let $\mathcal{A} = \{1, 2\}$ and $\mathcal{B} = \{a, b, c\}$, one relation would be to map 1 to all the vowels and 2 to all of the consonants. Then

$$\mathcal{R} = \{(1, a), (2, b), (2, c)\}$$

215

The definitions of a 1-1 relation and of an onto relation look identical to those for functions. Composition of relations is also defined identically. The definition of an inverse relation is *not* the same.

We require a function to be 1-1 and onto in order to have an inverse. *Every* relation has an inverse.

I'm not 1-1 onto; I have no inverse.

I'm not 1-1 onto but I *have* an inverse.

Definition *Inverse Relation*

Let $\mathcal{R}$ be a relation between the sets $\mathcal{A}$ and $\mathcal{B}$. The *inverse relation* of $\mathcal{R}$ is denoted $\mathcal{R}^{-1}$ and is a subset of the Cartesian product $\mathcal{B} \times \mathcal{A}$ for which

$$\mathcal{R}^{-1} = \{(b, a) \in \mathcal{B} \times \mathcal{A} \mid (a, b) \in \mathcal{R}\}$$

The definition says that the inverse relation is created by flipping the entries in each ordered pair.

That seems easy. I should be able to make up an example. I will let $\mathcal{A} = \{a, b, c, d\}$ and $\mathcal{B} = \{1, 2, 3\}$.
If $\mathcal{R} = \{(a, 2), (b, 1), (c, 3), (d, 2)\}$ then $\mathcal{R}^{-1} = \{(1, b), (2, a), (2, d), (3, c)\}$.

Both relations are onto. $\mathcal{R}$ is a function but it is not 1-1, so it has no inverse as a function. $\mathcal{R}^{-1}$ is *not* a function but it *is* a 1-1 relation.

That is a good example. You do seem to understand inverse relations.

There are four additional definitions for today. They describe some optional properties that relations between a set and itself might have (relations on $\mathcal{A} \times \mathcal{A}$).

Definition *Reflexive; Symmetric; Transitive*

Let $\mathcal{R}$ be a relation on a set, $\mathcal{A}$.

- $\mathcal{R}$ is *reflexive* if and only if $(x, x) \in \mathcal{R}$ for *all* $x \in \mathcal{A}$.
- $\mathcal{R}$ is *symmetric* if and only if for all $x, y \in \mathcal{A}$, $(x, y) \in \mathcal{R}$ implies $(y, x) \in \mathcal{R}$.
- $\mathcal{R}$ is *transitive* if and only if for all $x, y, z \in \mathcal{A}$, $((x, y) \in \mathcal{R}) \wedge ((y, z) \in \mathcal{R})$ implies $(x, z) \in \mathcal{R}$.

I have another definition: a *reflective* relation is one that thinks deeply about itself. (^_^)

Here are some examples.

$$\mathcal{R}_1 = \{(x, y) \in \mathbb{R} \times \mathbb{R} \mid x < y\}$$
$$\mathcal{R}_2 = \{(x, y) \in \mathbb{Z} \times \mathbb{Z} \mid x \text{ divides } y\}$$
$$\mathcal{R}_3 = \{(x, y) \in \mathbb{R} \times \mathbb{R} \mid \lceil x \rceil = \lceil y \rceil\}$$

(See page 54 for the definition of the ceiling function $\lceil x \rceil$.)

So, Lily, which of these are reflexive, symmetric, or transitive?

Well, $\mathcal{R}_1$ is not reflexive because $4 \not< 4$. It is not symmetric because $4 < 6$ but $6 \not< 4$. It *is* transitive: if $x < y$ and $y < z$, then it is always true that $x < z$.

$\mathcal{R}_2$ *is* reflexive because every integer divides itself. It is not symmetric because even though 4 divides 8, 8 does not divide 4. It is transitive since if x divides y, then all the prime factors of x are in y. But then if y divides z, all the prime factors of x that are sitting in y must also be in z.

$\mathcal{R}_3$ is reflexive, symmetric, and transitive. I think it is pretty easy to see that all three are true.

You are correct. That leads us to the last new idea for today. Relations that have all three properties are of special interest.

Definition *Equivalence Relation*

Let $\mathcal{R}$ be a relation on a set, $\mathcal{A}$. If $\mathcal{R}$ is reflexive, symmetric, and transitive, then it is called an *equivalence relation*.

An equivalence relations on a set $\mathcal{A}$ will partition $\mathcal{A}$ into a collection of *equivalence classes*.

The equivalence classes in $\mathcal{R}_3$ are the sets $(n - 1, n]$ for $n \in \mathbb{Z}$. For example, all the numbers in $(4, 5]$ round up to 5.

A really important example of an equivalence relation is congruence mod n. We can define the relation as

$$\mathcal{R} = \{(a, b) \in \mathbb{Z} \times \mathbb{Z} \mid a \equiv b \bmod n\}$$

We can choose a representative for each of the equivalence classes. For example, all the integers whose remainder on division by n is 2 would be in the same equivalence class as 2. We could denote that equivalence class as [2]. The following classes partition $\mathbb{Z}$: [0], [1], ..., [n − 1].

We can then define

$$\mathbb{Z}_n = \{[0], [1], [2], \ldots, [n - 1]\}$$

We can even define an addition and multiplication as

$$[a] + [b] = [(a + b) \bmod n]$$
$$[a] \cdot [b] = [(a \cdot b) \bmod n]$$

For $\mathbb{Z}_5$ we would have the following addition table.

+	[0]	[1]	[2]	[3]	[4]
[0]	[0]	[1]	[2]	[3]	[4]
[1]	[1]	[2]	[3]	[4]	[0]
[2]	[2]	[3]	[4]	[0]	[1]
[3]	[3]	[4]	[0]	[1]	[2]
[4]	[4]	[0]	[1]	[2]	[3]

For example,

$$[2] + [4] = [(2 + 4) \bmod 5]$$
$$= [6 \bmod 5]$$
$$= [1].$$

If n is a prime, we can even mimic pretty much all of the properties of the real numbers. For example, in $\mathbb{Z}_5$, every element has a multiplicative inverse. Consider the element [2]. Note that $[2] \cdot [3] = [(2 \cdot 3) \bmod 5] = [1]$. So [2] and [3] are multiplicative inverses.

I could say much more about this example, but that would take us beyond what we need for now.

That's ok. I am near my limit of new material for today.

I should mention that relations play a significant role in a subfield of computer science. Currently, the most common kind of database is a *relational database*. This model of a database uses relations as the fundamental building block. The rich mathematical theory of relations was used to validate the integrity and efficiency of the model.

The definition of a relation needs to be extended to Cartesian products of more than two sets. For example, you might have a set of id numbers, names, and grades. The database would store a table whose rows are relations (but often called *tuples*).

ID × Name × Grade

A small table in the database might contain:

ID	Name	Grade
414	Albert Alligator	D
228	Pogo Possum	A
397	Churchy LaFemme	C
860	Howland Owl	B

The labels at the top would not be part of the table.

There is much more to relational databases than what I have mentioned.

Well, I certainly learned things about functions I had not seen before. And relations are completely new. I will think about them over the next week.

Definitions

Cartesian Product Let A and B be sets. The *Cartesian product $A \times B$* is the set of all ordered pairs, (a, b), with first coordinate a a member of A and second coordinate b a member of B. That is,

$$A \times B = \{(a, b) \mid a \in A, b \in B\}$$

Function; Domain; Codomain; Range A *function* from the nonempty set $\mathcal{D}$ into the nonempty set $\mathcal{R}$ is a subset, $\mathcal{F}$, of the Cartesian product $\mathcal{D} \times \mathcal{R}$ such that every element of $\mathcal{D}$ appears as the first coordinate in one and only one ordered pair in $\mathcal{F}$.

The set, $\mathcal{D}$, is called the *domain* and the set, $\mathcal{R}$, is called the *codomain*. The *range* of the function is the subset of $\mathcal{R}$ consisting of elements that actually appear in the right-hand side of at least one ordered pair in $\mathcal{F}$.

A function can also be thought of as a mapping from the domain to the range such that no element of the domain gets mapped to more than one element of the range.

Onto and One-to-One Functions Let $\mathcal{F}$ be a function from $\mathcal{D}$ into $\mathcal{R}$. Then $\mathcal{F}$ is called *onto* if every element of $\mathcal{R}$ appears as a second coordinate in at least one ordered pair in $\mathcal{F}$. If no element of $\mathcal{R}$ appears as a second coordinate in more than one ordered pair in $\mathcal{F}$, then $\mathcal{F}$ is called *one-to-one*. (One-to-one is often abbreviated as 1-1).

Inverse Function Let $\mathcal{F}$ be a one-to-one and onto function with domain $\mathcal{D}_{\mathcal{F}}$ and range $\mathcal{R}_{\mathcal{F}}$. A function, $\mathcal{G}$, whose domain is $\mathcal{R}_{\mathcal{F}}$ and whose range is $\mathcal{D}_{\mathcal{F}}$ is called the *inverse of $\mathcal{F}$* if the following conditions hold:
- If $(x, y) \in \mathcal{F}$, then $(y, x) \in \mathcal{G}$.
- If $(y, x) \in \mathcal{G}$, then $(x, y) \in \mathcal{F}$.

Composition of Functions Let $\mathcal{F}$ be a function whose domain is $\mathcal{X}$ and whose range is $\mathcal{Y}$. Let $\mathcal{G}$ be a function whose domain is $\mathcal{Y}$ and whose range is $\mathcal{Z}$. The composition of $\mathcal{G}$ and $\mathcal{F}$ is denoted by $\mathcal{G} \circ \mathcal{F}$ and is defined by

$$\mathcal{G} \circ \mathcal{F} = \{(x, z) \mid \exists y \in \mathcal{Y}$$

$$\text{with } (x, y) \in \mathcal{F} \text{ and } (y, z) \in \mathcal{G}\}$$

Relation A *relation* between the set A and the set B is a subset, R, of the Cartesian product $A \times B$.

If $(a, b) \in R$, it is common to write aRb and to say that a is related to b. If $A = B$, the relation is said to be a *relation on* A.

The set A is called the *domain* of the relation and the set B is called the *codomain*.

Inverse Relation Let R be a relation between the sets A and B. The *inverse relation* of R is denoted R^{-1} and is a subset of the Cartesian product $B \times A$ for which

$$R^{-1} = \{(b, a) \in B \times A \mid (a, b) \in R\}$$

Reflexive; Symmetric; Transitive Let R be a relation on a set, A.

- R is *reflexive* if and only if $(x, x) \in R$ for *all* $x \in A$.
- R is *symmetric* if and only if for all $x, y \in A$, $(x, y) \in R$ implies that $(y, x) \in R$.
- R is *transitive* if and only if for all $x, y, z \in A$, $((x, y) \in R) \wedge ((y, z) \in R)$ implies $(x, z) \in R$.

Equivalence Relation Let R be a relation on a set, A. If R is reflexive, symmetric, and transitive, then it is called an *equivalence relation*.

Relational Database A model of a database that uses relations as the fundamental building block.

Exercises

Solutions can be found at `http://www.mathcs.bethel.edu/~gossett/DMGN/`.

1. Find the Cartesian products of the following sets.

    (a) $A = \{a, b, c\}$ $B = \{1, 2, 3\}$

    (b) $A = \{\flat, \sharp\}$ $B = \{£, ¥\}$

    (c) $A = \{a, b, c\}$ $B = \{(1, 2), (3, 4)\}$

2. Let $\mathcal{D} = \{w, x, y, z\}$ and $\mathcal{R} = \{4, 5, 6, 7\}$. Determine which of the following are

    - functions
    - onto
    - one-to-one

    Give reasons for your answers.

    (a) $\mathcal{E} = \{(w, 4), (x, 5), (y, 4), (z, 7)\}$

    (b) $\mathcal{F} = \{(w, 4), (x, 5), (y, 6), (w, 5), (z, 7)\}$

    (c) $\mathcal{G} = \{(w, 5), (x, 6), (y, 7), (z, 4)\}$

    (d) $\mathcal{H} = \{(w, 4), (x, 4), (y, 4), (z, 4)\}$

3. The following all represent functions. In each case, determine whether an inverse function exists. If one does exist, find it.

    (a) $D = \{a, b, c\}$ $R = \{1, 2, 3\}$ $\mathcal{F} = \{(a, 2), (b, 3), (c, 1)\}$

    (b) $D = \{a, b, c\}$ $R = \{1, 2, 3\}$ $\mathcal{F} = \{(a, 2), (b, 2), (c, 1)\}$

    (c) $D = \{a, b, c, d\}$ $R = \{1, 2, 3\}$ $\mathcal{F} = \{(a, 2), (b, 3), (c, 1), (d, 2)\}$

    (d) $D = \{a, b, c\}$ $R = \{1, 2, 3, 4\}$ $\mathcal{F} = \{(a, 2), (b, 3), (c, 1)\}$

4. Let $X = \{a, b, c\}$ $Y = \{1, 2, 3\}$ $Z = \{x, y, z\}$. Find the composition $\mathcal{F} \times \mathcal{G}$ for the following pairs of functions.

    (a) $\mathcal{F} = \{(a, 2), (b, 2), (c, 1)\}$ $\mathcal{G} = \{(1, y), (2, z), (3, y)\}$

    (b) $\mathcal{F} = \{(a, 3), (b, 2), (c, 1)\}$ $\mathcal{G} = \{(1, z), (2, x), (3, y)\}$

    (c) $\mathcal{F} = \{(a, 2), (b, 2), (c, 2)\}$ $\mathcal{G} = \{(1, y), (2, x), (3, z)\}$

    (d) $\mathcal{F} = \{(a, 1), (b, 2), (c, 3)\}$ $\mathcal{G} = \{(1, x), (2, y), (3, z)\}$

5. Let $\mathcal{D} = \{w, x, y, z\}$ and $\mathcal{R} = \{4, 5, 6, 7\}$. For each of the following relations

   - determine if it is onto
   - determine if it is one-to-one
   - find its inverse

   (a) $\mathcal{E} = \{(w, 4), (x, 5), (y, 4), (z, 7)\}$

   (b) $\mathcal{F} = \{(w, 4), (x, 5), (y, 6), (w, 5), (z, 7)\}$

   (c) $\mathcal{G} = \{(w, 5), (x, 6), (y, 6), (y, 7), (z, 4), (z, 5)\}$

   (d) $\mathcal{H} = \{(w, 4), (x, 4), (y, 4), (z, 4)\}$

6. Let $X = \{a, b, c, d\}$ For each of the following relations on X, determine if it is

   - reflexive
   - symmetric
   - transitive
   - an equivalence relation (if so, list the equivalence classes)

   Give reasons for your answers.

   (a) $A = \{(a, a), (a, b), (b, a), (b, b), (b, c), (c, d), (d, d), (d, c)\}$

   (b) $B = \{(a, a), (a, b), (b, a), (b, b), (b, c), (c, a),$
   $(c, b), (c, c), (c, d), (d, c), (d, d)\}$

   (c) $C = \{(a, b), (b, c), (c, d)\}$

   (d) $D = \{(a, a), (b, b), (c, c), (d, d)\}$

   (e) $\mathcal{E} = \{(a, a), (a, b), (b, a), (b, b), (c, c)(c, d), (d, c), (d, d)\}$

   (f) $\mathcal{F} = \{(a, a), (a, b), (a, c), (a, d), (b, a), (b, b), (b, c), (b, d),$
   $(c, a), (c, b), (c, c)(c, d), (d, a), (d, b), (d, c), (d, d)\}$

7. For each of the following relations on A, determine if it is

   - reflexive
   - symmetric
   - transitive
   - an equivalence relation (if so, list the equivalence classes)

   Give reasons for your answers.

   (a) $A = \mathbb{R}$ $\mathcal{R}_1 = \{(x, y) \in \mathbb{R} \times \mathbb{R} \mid x \leq y\}$

   (b) $A = \mathbb{R}$ $\mathcal{R}_2 = \{(x, y) \in \mathbb{R} \times \mathbb{R} \mid x - y = 4\}$

   (c) A is the set of all living people.
   $\mathcal{R}_3 = \{(x, y) \in A \times A \mid x \text{ and } y \text{ are cousins}\}$

Chapter 15

Hi Isolde. I know today is our final session, so I am feeling quite sad. I don't know how well I will be able to concentrate on the topic you have chosen.

I don't plan to introduce a new topic today. Instead, I thought it would be helpful to look back on what we have both learned.

Before we start, I wanted to tell you how much I have enjoyed our time together. I have appreciated your enthusiasm and your ability to ask good questions. In addition, I like you very much as a person.

what are you doing grandma?

I am reminising about the past.

Isolde, I am going to miss our time together! I have had so much fun and learned so many interesting things. Can we still get together sometimes?

Yes, I would love to meet with you again. Maybe we can get together for dinner and then exchange math puzzles.

To get started on our time of reflection, let me tell you a story from my first year as a university student. When I first started college, I was intimidated when I learned some of my courses would have comprehensive final exams.

During the Spring Semester, one of my math professors told us why she was giving a comprehensive final exam. She said that looking back at the entire semester provides several benefits. The first is that the additional review helps to move the material into long term memory rather than having it disappear soon after the exam. Another benefit is to make a deliberate inventory of what has actually been learned during the course. (I was amazed at how much I had learned in that course. I also noticed that some of the topics that were really difficult for me when I first encountered them were now things I understood pretty well.) The final advantage my teacher mentioned was that it gave us an opportunity to consider connections between her course and other math courses we had taken in the past.

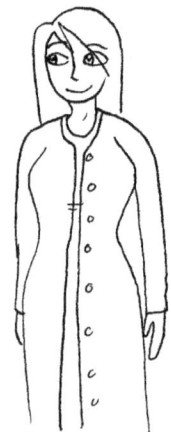

We won't have a final exam, but I do have some questions to help you reflect on our time together. First, how would you describe discrete mathematics to a friend who hasn't studied the subject?

Lily thinks for a few minutes.

We studied lots of topics, so it's not easy to see what ties them all together. Two things occur to me. First, there was an emphasis on understanding the reasoning behind the topics and ideas. At times there were proofs (and you said that a university course in Discrete Mathematics would contain a whole lot of proofs). Another common aspect is that most of the topics were concerned with finite collections of things. For example, the vertices and lines in a graph, the items we counted, the steps in an algorithm, and the states in a finite-state machine. I'm sure I missed many other examples.

Have you noticed any connections between what we have discussed and any of your previous learning about mathematics?

I have seen a little bit about sets, but not as much as you presented. Some of the material about elementary number theory was familiar but I never thought about it as formally as we did. The counting material went way beyond what I imagined would be part of that topic. The proof techniques were completely new, as were several other topics.

Some of the study suggestions were new, but some reinforced things I was already doing. You just gave me a way to describe why they seemed to be useful ways to approach studying.

Can you give an example of one suggestion that reinforced what you were already doing?

Yes. I was already memorizing things as I studied. I did so because I thought that was what my teachers wanted on exams—repeat back what they had told me. You got me thinking about how having one thing memorized helps me to notice how it might apply to a different topic. For example, I would have had a harder time understanding the reasoning behind indirect proof if we had not first studied logic. Knowing that an implication and its contrapositive are logically equivalent is key to understanding indirect proof.

Here are some other questions to ponder on your own:

- How have you grown as a learner during our time together?

- What topics did you find most useful?

- Which topics were the most interesting?

- Is there a topic about which you would like to learn more?

- Is there a study strategy that you would like to develop more fully in the future?

what have I learned?

Can I ask what you most enjoyed, Isolde?

I really like the counting topics. I also had fun with the puzzles you brought.

Speaking of puzzles, let's end with a final puzzle.

Suppose you have 80 marbles and a balance scale. One marble weighs more than the rest. The other 79 all weigh the same (or are so similar that the scale cannot distinguish any difference). Assume the scale can hold all the marbles at once. How can you find the heavy marble in only 4 weighings?

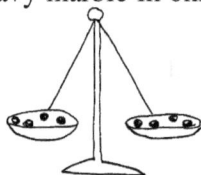

Can I get a hint?

Try to solve a few smaller versions of the problem.

Ok. If I have 4 marbles, I could do it in two weighings. Place 2 marbles in each side of the scale. Then keep the two on the tray that goes down. Now put one of these in each tray and keep the one that makes the tray go down. That is the heavy marble.

If I start with 5 marbles. I could start by putting 2 marbles in each tray. If they balance, then the leftover marble is the heavy one. Otherwise, I would follow the four-marble strategy. So I can still get by with at most 2 weighings.

With 6 marbles, I can start by weighing 4 of them. If they balance, then weigh the other two. If they don't balance, then follow the four-marble strategy. Again I can do it in 2 weighings.

Can you think of a different way to solve the problem with 6 marbles?

Lily thinks for a while.

I suppose I could split them into two groups of 3 marbles and weigh them. I would then take the heavier bunch and weight two of the marbles. If they balance, the third marble is the heavy one. Otherwise, I would select the marble that is on the lower tray.

Good. Are you ready now to try solving the full 80 marble problem?

I suppose I could try the strategy of splitting the marbles into two groups. I would have 40 on each tray for the first round, then 20 per tray for round two. Rounds three and four would have 10 and 5 per tray, leaving me with two more rounds to guarantee a solution. That is six rounds, so this strategy fails.

Plan B is to try the strategy where I

split the marbles into three groups and weigh two of them. My initial split would need to be in groups of sizes 26, 27, and 27. In round one I could put the groups of size 27 on the scale. I can make an outline of the steps.

Lily's outline is on the next page. Try to produce it yourself before turning the page.

Congratulations Lily. Your strategy solves the problem.

This is the end of our tutoring sessions. I want to thank you again for being such a diligent student. I enjoyed our time much more than I expected.

I hope you keep your enthusiasm for learning new things. Learning takes work, but it can be very rewarding.

Thank you for all the time you spent preparing for our sessions. You are like a model big sister to me. I do hope we can stay in touch.

Thank You for reading this book!

Lily's Solution to the 80 marbles problem

1. Subdivide into groups of size 27, 27, and 26. Weigh the groups of size 27. If the scale tips, keep the lower group. If the scale balances, keep the group with 26 marbles.

2. I now have either 26 or 27 marbles.

   27 marbles Subdivide into three groups of 9 marbles each. Weigh any two groups. If the scale tips, keep the lower group. If the scale balances, keep the group that wasn't weighed.

   26 marbles Subdivide into groups of size 9, 9, and 8. Weigh the groups of size 9. If the scale tips, keep the lower group. If the scale balances, keep the group with 8 marbles.

3. At this stage, I have either 8 or 9 marbles.

   9 marbles Subdivide into three groups of 3 marbles each. Weigh any two groups. If the scale tips, keep the lower group. If the scale balances, keep the group that wasn't weighed.

   8 marbles Subdivide into groups of size 3, 3, and 2. Weigh the groups of size 3. If the scale tips, keep the lower group. If the scale balances, keep the group with 2 marbles.

4. I now have either 2 or 3 marbles.

   3 marbles Weigh any two of the marbles. If the scale tips, the lower marble is the one I seek. If the scale balances, the marble that wasn't weighed is the one I seek.

   2 marbles Weigh the marbles. The scale will tip. The lower marble is the one I seek.

Appendix 1: Log Function Review

You have probably encountered logarithmic functions in one or more of your math classes. Here is a brief review to make sure the ideas are fresh for future reference.

Definition *Logarithmic Functions*

The *logarithmic function with base b*, denoted $\log_b(x)$, is defined by the relationship

$\log_b(x) = y$ for $x > 0$ if and only if $b^y = x$,   where $0 < b$ and $b \neq 1$

The most commonly used logarithmic functions are

common logs $\log_{10}(x)$, usually denoted $\log(x)$ on calculators

natural logs $\log_e(x) = \int_1^x \frac{dt}{t}$, usually denoted $\ln(x)$

base 2 logs $\log_2(x)$, the prime candidate in discrete mathematics

---

Theorem *Properties of Logarithmic Functions*

- $\log_b(1) = 0$

- $\log_b(b) = 1$

- $\log_b(x^n) = n \log_b(x)$ for any real number n and $x > 0$

- $\log_b(xy) = \log_b(x) + \log_b(y)$ for $x > 0$ and $y > 0$

- $\log_b(\frac{x}{y}) = \log_b(x) - \log_b(y)$ for $x > 0$ and $y > 0$

Note well: There are no theorems for simplifying $\log_b(x + y)$ or $\log_b(x - y)$.

Theorem *Change of Base Formula*

For all legal bases a and b and all $x > 0$,

$$\log_b(x) = \frac{\log_a(x)}{\log_a(b)}$$

Appendix 2: Summation Notation Review

Summation notation is just a convenient abbreviation for a long sum, usually involving subscripted variables. We use an index (i in the following definition) to indicate how to cycle through the variables, adding each new variable to the sum. The Greek letter Sigma, Σ, is used to denote the sum. Below and above the symbol we indicate the starting and ending values of the index.

Definition *Summation Notation*

$$\sum_{i=k}^{n} a_i = a_k + a_{k+1} + a_{k+2} + \cdots + a_{n-1} + a_n$$

If $k = 3$ and $n = 7$, then $\sum_{i=k}^{n} a_i = a_3 + a_4 + a_5 + a_6 + a_7$. The choice of index variable does not change the sum: $\sum_{i=k}^{n} a_i = \sum_{j=k}^{n} a_j$. Changing the starting or ending value for the index *does* change the sum.

The symbol c represents, in the properties below, a number that does not change with the index. You will improve your understanding if you make up simple examples to test the properties. For example, $\sum_{i=1}^{4}(c \cdot a_i) = c \cdot \sum_{i=1}^{4} a_i$ is merely an extended version of the distributive property of the real numbers:

$$ca_1 + ca_2 + ca_3 + ca_4 = c(a_1 + a_2 + a_3 + a_4)$$

1. $\displaystyle\sum_{i=1}^{n} c = nc$

2. $\displaystyle\sum_{i=k}^{n}(a_i + b_i) = \sum_{i=k}^{n} a_i + \sum_{i=k}^{n} b_i$ The parentheses on the left are required.

3. $\displaystyle\sum_{i=k}^{n}(a_i - b_i) = \sum_{i=k}^{n} a_i - \sum_{i=k}^{n} b_i$

4. $\displaystyle\sum_{i=k}^{n}(c \cdot a_i) = c \cdot \sum_{i=k}^{n} a_i$

References and Credits

Courtesy of Eric Gossett

Funding This project was partially funded by a Faculty Development Grant from Bethel University, in St. Paul, Minnesota. The author is deeply grateful for the support.

Mathematical Content Unless otherwise credited, mathematical content can be found in *Discrete Mathematics With Proof*, 2nd edition, by Eric Gossett, John Wiley & Sons, Inc, © 2009, ISBN: 978-0-470-45793-1.

Definitions and Terms Terms and definitions throughout text are from Discrete Mathematics with Proof, Eric Gossett. Copyright © 2009 by John Wiley & Sons, Inc. Reproduced with permission of John Wiley & Sons, Inc.

Acknowledgements I would like to thank my wife, Florence, for reading the first draft. She made numerous helpful suggestions and corrections.

Page v The equation on the board is known as Pascal's Theorem. You can learn a bit more about the theorem on page 159.

Page vi *Make It Stick: the Science of Successful Learning*, by Brown, Rodediger, McDaniel, The Belknap Press of Harvard University Press, 2014, ISBN: 978-0-674-72901-8, page 15

Page 32 *Make It Stick: the Science of Successful Learning*, by Brown, Rodediger, McDaniel, The Belknap Press of Harvard University Press, 2014, ISBN: 978-0-674-72901-8, page 92

Page 33 Rodin's *The Thinker* is the model for the chapter title image.

Page 34 *The Mikado* is a comic opera by Gilbert and Sullivan.
`https://en.wikipedia.org/wiki/The_Mikado`

Page 45 The cemetery puzzle is from
`http://io9.com/to-solve-this-riddle-look-to-your-family-1654578754`

Page 46 Some of the supporting reasons for memorization come from conversations with my colleagues Jay Rasmussen (Education) and Scott Brown (Mathematics Education).

Page 61 The apple puzzle is from
`http://www.mathsisfun.com/puzzles/apples-and-friends-solution.html`

Page 61 *Make It Stick: the Science of Successful Learning*, by Brown, Rodediger, McDaniel, The Belknap Press of Harvard University Press, 2014, ISBN: 978-0-674-72901-8, page 63

Page 75 Inspiration for the chapter title image is from
`https://legendaryarchive.files.wordpress.com/2015/04/selkie.jpg`

Page 86 The candy bar complete induction example was found at
`https://www.cs.cmu.edu/~adamchik/21-127/lectures/induction_2_print.pdf`
from a lecture by Victor Adamchik.

Page 89 The model for the chapter title image is from
`http://illuminations.nctm.org/Lesson.aspx?id=655`

Page 90 Information about the Lo Shu magic square was found on these web pages:
`http://illuminations.nctm.org/Lesson.aspx?id=655`
`http://mysteriouswritings.com/`
`  sacred-pattern-from-the-lo-river-the-first-magic-square/`
`https://en.wikipedia.org/wiki/Lo_Shu_Square`
`http://www.smart-kit.com/s908/very-old-chinese-book-puzzle/`

Page 92 The 4 by 4 magic square (and related information) is from
`http://www.math.wichita.edu/~richardson/mathematics/magic%`
`20squares/4th-ordermagicsquares.html`

Page 103 The Gerbil Welfare Committee is the invention of Tristan Hoppe.

Page 116 *The Mythical Man-Month*, anniversary edition, by Frederick P. Brooks, Jr., Addison-Wesley Longman, 1995, ISBN: 0-201-83595-9

Page 130 *Make It Stick: the Science of Successful Learning*, by Brown, Rodediger, McDaniel, The Belknap Press of Harvard University Press, 2014, ISBN: 978-0-674-72901-8, page 43

Page 133 You can learn more about Georg Cantor at
`http://www-history.mcs.st-andrews.ac.uk/Biographies/Cantor.html`

Page 133 "Scathingly brilliant": see *The Trouble with Angels*
`http://www.imdb.com/title/tt0061122/reviews`

Page 136 The "2 years vs 10 seconds" assertion. *Discrete Mathematics With Proof*, 2nd edition, Eric Gossett, Wiley, 2009, ISBN: 978-0-470-45793-1, page 188.

Page 142 The complexity measures for binary search can be found in *Discrete Mathematics With Proof*, 2nd edition, by Eric Gossett, Wiley, 2009, ISBN: 978-0-470-45793-1, pages 197 and 689.

Page 146 *The Mutilated Chessboard* puzzle is from *The Scientific American Book of Mathemtatical Puzzles & Diversions*, by Martin Gardner, Simon and Schuster, 1959, page 24. ISBN for a newer printing by The University of Chicago Press: 978-0226282534

Page 149 The logo is an example of a Persian Rug design, created using a Java applet created by the author. The inspiration is the article *Persian Recursion* by Anne M. Burns, *Mathematics*, 1997, volume 70, number 3, pages 196-199.

Page 162 The rope-around-the-globe problem has been around for a long time. It can be found in multiple sources. I don't know the original source.

Page 163 Estimates for the earth's circumference at the equator can be found at the following web sites.

`http://geography.about.com/library/faq/blqzcircumference.htm`
`https://en.wikipedia.org/wiki/Earth`

Page 165 The finite-state machine is inspired by Peter Pan's reply when Wendy asks him where he lived. "Second to the right," said Peter, "and then straight on till morning." *Peter Pan*, by J. M. Barrie, Book-of-the-Month Club, 2003, page 27. Originally published in 1911.

Page 171 The blue police box is the Tardis (Time And Relative Dimension In Space). It is, of course, Doctor Who's time and space machine "bigger on the inside than on the outside." `https://en.wikipedia.org/wiki/TARDIS`

Page 176 Careercast.com ranked actuary as the best job and mathematician as the third best job for 2015.
`http://www.careercast.com/jobs-rated/best-jobs-2015`
`http://www.careercast.com/slide/best-jobs-2015-no-3-mathematician`

Page 176 One of my former students, Kaysee Maas, has actually worked on a medical laboratory location project while in graduate school.

Page 177 *Perplexing Puzzles and Tantalizing Teasers*, by Martin Gardner, illustrations by Laszlo Kubinyi, Arch, 1971, page 103. Current ISBN: 978-0486256375

Page 179 The top image is the dodecahedron, a regular polygon with 12 faces, 30 edges, and 20 vertices. The bottom image is a planar embedding of the dodecahedron – one face has been removed and the polyhedron has been stretched and flattened. The images were made using Mathematica 10.2.

Page 182 You can find more on the Seven Bridges of Königsberg at

`https://en.wikipedia.org/wiki/Seven_Bridges_of_K%C3%B6nigsberg`

Page 182 The map of the Pregel river and the Königsberg bridges is from "Konigsberg bridges" by Bogdan Giuşcă - Public domain (PD), based on the image. Licensed under CC BY-SA 3.0 via Wikimedia Commons -

`https://commons.wikimedia.org/wiki/File:Konigsberg_bridges.png#`
`  /media/File:Konigsberg_bridges.png`

Page 182 The sketch of Leonard Euler was inspired by

`http://www-history.mcs.st-and.ac.uk/BigPictures/Euler_8.jpeg`

Page 192 *Number Stories Of Long Ago*, by David Eugene Smith, 1962, The National Council of Teachers of Mathematics, page 119.

Page 208 The account of Solomon and the two women can be found in the Bible in 1 Kings chapter 3, verses 16 to 28.

Page 209 The fair division algorithm is outlined at

`http://www.colorado.edu/education/DMP/fair_division.html`

Page 218 The names come from the comic strip *Pogo*, by Walt Kelly. The strip was in syndication from 1949 to 1973. The strip was the origin of the famous statement "we have met the enemy and he is us."

Page 224 Some of the ideas in this chapter were a result of a conversation with my colleague Scott Brown (Mathematics Education).

Page 231 Two of the author's grandchildren at the Oliver Kelley Farm, run by the Minnesota Historical Society.

Index

incident, 181, 187
inclusive or, 35
independent, 96
independent tasks principle, 97, 107, 127
indirect proof, 66, 68, 73
induction
    complete induction, 85, 87
    mathematical induction, 81, 87
inference, 66
integers, 48
integers mod n, 53
interior node, 197
intersection, 13, 18
inverse
    relation, 216, 220
inverse function, 214, 219
involution, 24, 29, 41
Irish step dance, 192
irrational numbers, 48

Jane Austen, 12

Königsberg, 182
Königsberg bridge problem, 182
key, 201
Kuratowski's Theorem, 184, 188

Law of Addition, 41
Law of Contradiction, 39, 41
Law of Double Negation, 41
Law of Hypothetical Syllogism, 66
law of simplification, 42
Law of the Excluded Middle, 39, 41
lcm, 51, 55
leaf node, 197
learning strategies
    brain processing during sleep, 62,
        114
    cramming, 61
    effort, 32
    effortful retrieval, 130

failure, 32
get enough sleep!, 114
instructors and tutors, 114
memorization, 46
proper prerequisites, 113
read the textbook carefully, 113
repeated retrieval, 130
review, 61, 62
start homework early, 113
time on task, 112
time-spaced retrieval, 130
when you are stuck, 113
working with classmates, 113, 114
least common multiple, 51, 55
lemma, 65, 72
leprechaun, 76
level, 197
LHRRWCC, 155, 158
Lichtenberg, Georg Christoph, 44
linear homogeneous recurrence relation
        with constant coefficients, 155,
        158
Lo river, 90
Lo shu square, 90
logarithms, 229
logic
    connectives
        AND, 35, 41
        biconditional, 37, 41
        implication, 37, 41
        NOT, 35, 41
        OR, 35, 41
    exclusive OR, 35
    inclusive OR, 35
    logically equivalent, 36
    proposition, 34
    statement, 34
    tautology, 37, 41
    truth table, 35, 41

www.ingramcontent.com/pod-product-compliance
Lightning Source LLC
Chambersburg PA
CBHW061405210326
41598CB00035B/6100